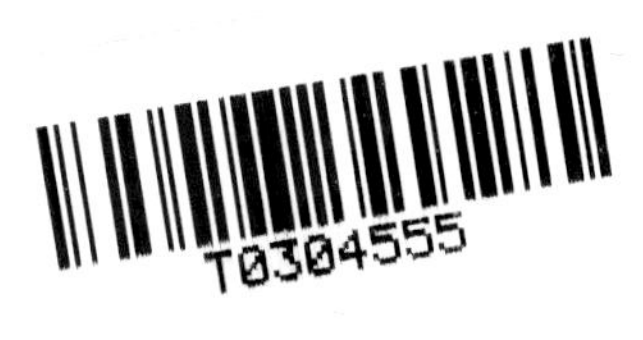
T0304555

# Nanophononics

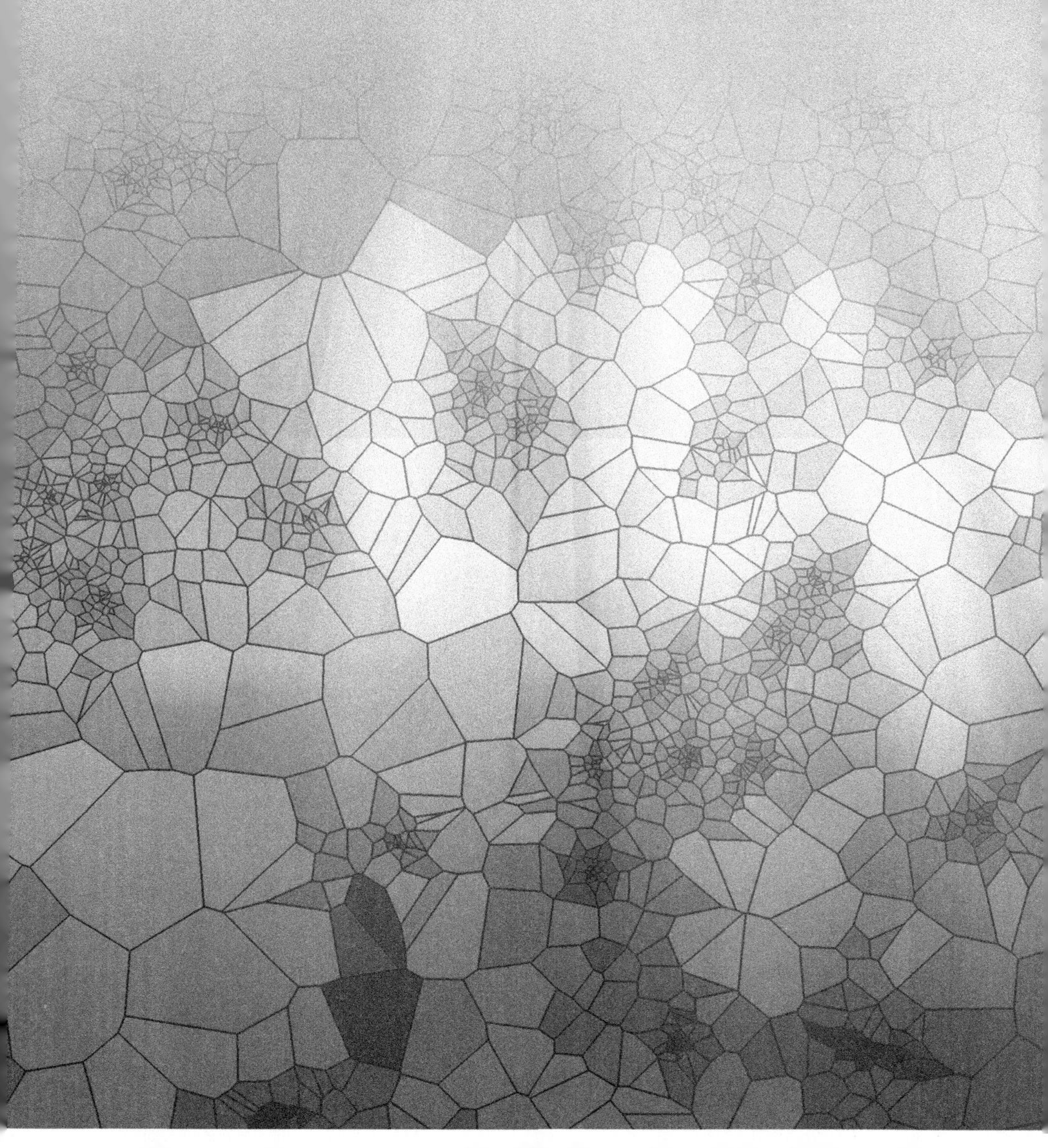

# Nanophononics

## Thermal Generation, Transport, and Conversion at the Nanoscale

edited by

**Zlatan Aksamija**

*Published by*

Pan Stanford Publishing Pte. Ltd.
Penthouse Level, Suntec Tower 3
8 Temasek Boulevard
Singapore 038988

Email: editorial@panstanford.com
Web: www.panstanford.com

**British Library Cataloguing-in-Publication Data**
A catalogue record for this book is available from the British Library.

**Nanophononics: Thermal Generation, Transport, and Conversion at the Nanoscale**



ISBN 978-981-4774-41-3 (Hardcover)
ISBN 978-1-315-10822-3 (eBook)

# Contents

# Preface

As the title implies, this book merges "phonons," quantized packets of lattice vibrations that are the primary carriers of heat in most semiconductors, with "nanoscale phenomena." The goal of this combination is to explore what happens to heat as we scale our nanoelectronic, optoelectronic, and energy devices down to feature sizes in the hundreds of, tens of, and even single nanometers. At such extreme scales, the thermal energy stored in and carried by phonons takes on new and emerging properties, such as ballistic transport and confinement, while modulating the coupling between electronic and thermal transport. Dissipation at the nanoscale, thus, becomes a new challenge but one rife with opportunities for discovery and improvement via nanoengineering.

This book grew out of a special session on nanophononics that I organized at the IEEE Nano conference in Rome, Italy. The session brought together speakers who were studying phonons in nanostructures, with a particular focus on those individuals who are at the cusp between electrical engineering, a core area traditionally represented at the IEEE Nano conference; mechanical engineering, as it has historically encompassed heat transfer; and materials science for its contributions to first-principle materials modeling. While research in all three of these areas was affected by dissipation and phonons, they were infrequently brought together in conferences.

After the conference, the authors of the chapters in this volume built off the work they presented at IEEE Nano and incorporated their most up-to-date findings and results. The fruit of their labors is this edited volume on nanophononics, which focuses on thermal effects in nanostructures, including the generation, transport, and conversion of heat at the nanoscale level. It covers semiconductor nanostructures, including the traditional group IV elements (e.g., Si, Ge, diamond), carbon allotropes (graphene and graphene nanoribbons), and emerging new materials like transition metal dichalcogenides (TMDCs). The volume could be roughly divided into four segments focusing on the main themes of this book: (i) phonon generation or heat dissipation, (ii) nanoscale phonon transport,

(iii) applications and devices (including thermoelectrics), and (iv) emerging materials (graphene or two-dimensional).

In the first theme, phonon generation through interactions with electrons in out-of-equilibrium conditions and devices and light (photovoltaics) is covered. Nanoscale phonon transport, ranging from diffusive to ballistic, and anharmonic decay are covered in the second theme, comprising Chapters 4 and 5. In addition, theoretical and numerical simulation methods, such as phonon Monte Carlo and first-principle calculations, feature prominently in the second part of the book. The third theme, applications and devices, focuses on heat dissipation and self-heating in nanoelectronics and thermoelectric energy conversion. The fourth theme focuses on emerging issues such as tandem photovoltaic–thermoelectric converters and ultranarrow graphene ribbons.

Overall, this book will provide researchers, graduate students, and practitioners with a solid reference on the role of phonon transport at the nanoscale and in a variety of applications. I extend my sincere gratitude to all the chapter authors for their generous contributions to this book and their patience while the book came together. This volume, however, is by no means comprehensive: nanophononics remains an active area of research where numerous scientists and engineers continue to make breakthroughs by understanding, designing, building, and testing new phononic crystals, thermal diodes, and a slew of other materials and devices that take advantage of the unique thermal properties of nanostructures. It is my hope that the book will spark many deep conversations across disciplinary boundaries, between theorists and experimentalists, and between materials scientists and device engineers.

**Zlatan Aksamija**

University of Massachusetts Amherst

2017

# Chapter 1

# Modeling Self-Heating Effects in Nanoscale Devices

**Katerina Raleva,[a] Abdul Rawoof Shaik,[b] Suleman Sami Qazi,[b] Robin Daugherty,[b] Akash Laturia,[b] Ben Kaczer,[c] Eric Bury,[c] and Dragica Vasileska[b]**

[a]*Sts. Cyril and Methodius University, Faculty of Electrical Engineering and ITs, Skopje, Macedonia, Greece*

[b]*Arizona State University, School of Electrical, Computer and Energy Engineering, Tempe, AZ, USA*

[c]*Imec, Kapeldreef 75, 3001 Heverlee, Belgium*

vasileska@asu.edu, catherin@feit.ukim.edu.mk

## 1.1 Introduction

In recent years, technology has advanced to fabricate integrated circuits (ICs) at 14 nm gate length commercially [1]. Fab industry giants, such as Intel, TSMC, Samsung, and Global Foundries, have plans to fabricate ICs at 10 nm by 2017. Samsung has already fabricated and tested 128 Mb SRAM in 10 nm [2] and is hoping to commercialize the process by the end of 2016. This aggressive scaling of technology is possible because of the advent of fin-shaped

*Nanophononics: Thermal Generation, Transport, and Conversion at the Nanoscale*
Edited by Zlatan Aksamija

ISBN 978-981-4774-41-3 (Hardcover), 978-1-315-10822-3 (eBook)
www.panstanford.com

field-effect transistors (FinFETs) and fully depleted silicon-on-insulator (FD-SOI) device technology.

The device miniaturization is not without problems, however. It results in large amounts of heat generated per unit volume [3]. Self-heating leads to overheating of the device, which degrades the device performance and also affects device reliability. A recent study on self-heating of 14 nm down to 7 nm silicon FinFETs shows that heat confinement in the Si channel increases by 20% and in the strained Ge channel by 57% [4]. This, in turn, results in 70 K and 100 K change in channel temperature in 14 nm and 7 nm field-effect transistors (FETs), respectively. Hence efficient heat removal methods are necessary to increase device performance and device reliability. A recent trend in peak efficiency versus power density of the switched capacitor power converters shows that the efficiency decreases as power density increases [5]. This implies that the efficiency of utilizing the electrical energy in logic operations is getting reduced as the power density is increasing.

The control of electrons and holes in semiconductors has resulted in great technological advancement in computing, signal processing, biomedical applications, etc. The control of photons in different materials has generated technological revolution in wireless communications, optical fiber applications, microwave applications, etc. These applications are significantly helpful in many day-to-day activities. Another particle similar to electrons and photons is phonon. It is a particle that carries heat energy in the materials. Like electrons and photons, with the control of phonons in materials new innovations are possible [6]. For example, acoustic diodes and thermal diodes are the diodes that pass sound and heat in a unidirectional way similar to an electronic diode. To understand how these devices operate and propose new innovations with these devices, the phonon transport problem should be studied and understood very well. Hence, the study of phonon transport in materials is a necessary step to make further advances in technology.

The present work focuses on understanding the heat transport in nanoscale electronic devices. In Section 1.2 we describe the mathematical model for incorporation of self-heating in FD-SOI devices and conventional metal-oxide-semiconductor field-effect transistors (MOSFETs) that has been used in this work. Section 1.3 is devoted to a brief description of simulation results, including:

- FD-SOI device technology generations (Section 1.3.1)

- Introduction of structures for reducing self-heating (dual-gate [DG] devices, silicon on diamond [SOD], silicon on AlN, etc.) (Section 1.3.2)
- Application of the model to a heater-sensor circuit in common-source and common-drain configuration that is used to uncover the temperature of the hot spot (Section 1.3.3)

Conclusive comments from this work are presented in Section 1.4, together with possible directions for future research.

## 1.2 Self-Heating

### 1.2.1 General Considerations

It is well-known that whenever a temperature gradient exists within a medium, thermal energy flows from the region of higher temperature to the one with a lower temperature. This phenomenon is known as heat conduction and is described by Fourier's law [7]:

$$\mathbf{q} = -\kappa \nabla T, \tag{1.1}$$

where $\mathbf{q}$ is the heat flux vector, $T$ is the local temperature, and $\kappa$ is the thermal conductivity. If it is assumed that the local thermal energy can be described by the temperature, one can write a continuity equation for energy that involves a change of the local energy in time with the divergence of the heat flux,

$$\nabla^2 T - \frac{1}{\alpha}\frac{dT}{dt} = -\frac{1}{\kappa} q_{\text{gen}}, \tag{1.2}$$

where $q_{\text{gen}}$ is the heat generated. Thermal diffusivity ($\alpha$) is related to thermal conductivity ($\kappa$), specific heat ($c$), and density ($\rho$) by

$$\alpha = \kappa/(\rho c). \tag{1.3a}$$

For steady-state problems, the heat conduction equation simplifies to

$$\kappa \nabla^2 T + q_{\text{gen}} = 0. \tag{1.3b}$$

In conventional electrothermal simulations, it is a common practice to associate the generated power density that appears in Eq. 1.3b with Joule heating due to the presence of an electrical current and the resistivity of the material, that is,

$$q_{\text{gen}} = \mathbf{J} \cdot \mathbf{E}, \tag{1.3c}$$

where **J** is the local current density and **E** is the local electric field, hence representing Joule heating as a local quantity. By introducing position-dependent thermal conductivity, the heat conduction Eq. 1.3b can be expressed as

$$\nabla \cdot (\kappa \nabla T) + \mathbf{J} \cdot \mathbf{E} = 0. \tag{1.4}$$

In commercial device simulators, the heat conduction equation is coupled to the Joule heating term with either the drift-diffusion or energy balance equations of the carriers. This then leads to the so-called nonisothermal drift-diffusion or energy balance models [8–10]. The coupling between the electron/hole transport in devices and the corresponding heat flow is achieved via temperature-dependent mobilities and diffusion coefficients in the corresponding expression for current in Eq. 1.4. Thus, on the one hand, the lattice temperature enters the expression for the local mobility value, which, in turn, affects the electrostatics and the current density in the device. On the other hand, lattice temperature affects the local Joule heating term, which, in turn, affects the lattice temperature profile. The electrical conduction and the heat flow equations are then self-consistently solved for the temperature and the electrostatic potential. A natural question then is, Is this novel self-consistent-solution model for electrical and thermal conduction within nanoscale devices valid when the ballistic, nonstationary transport dominates within these devices and the mobility is no longer expressed by the classical picture [11]?

Many research groups have demonstrated that in nanoscale thermal transport, the use of simple Fourier's law for heat conduction is inappropriate but rather the phonon Boltzmann transport equation must be solved directly to calculate accurately, for example, thermal conductivity within thin films where boundary scattering plays significant role in thermal conductivity degradation. This is needed in order to explain the already existing experimental data [12]. The length scale at which Fourier's heat law needs to be replaced with a direct solution of the phonon Boltzmann transport equation, to explain heat transport through a nanoscale medium, is shown schematically in Fig. 1.1.

A direct solution of the phonon Boltzmann transport equation is a very difficult task for several reasons. First, it is difficult to express the anharmonic phonon decay processes mathematically. Second, one has to solve phonon Boltzmann equations for each individual mode of the acoustic and optical branches. A few attempts for solving the problem using the relaxation time approximation (RTA) have been made by Narumanchi et al. [13]. If electrons and holes are included in the picture with their corresponding Boltzmann transport equations, the solution of a coupled set of equations, comprising the electron-hole-phonon system, becomes a formidable task even for modern-day high-performance computing systems.

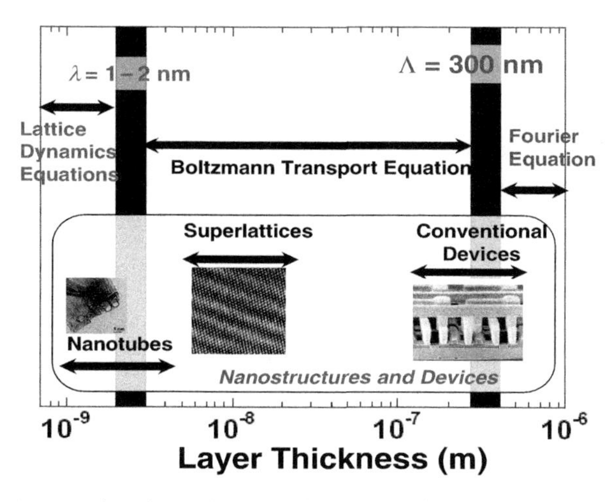


**Figure 1.1** Length scales and relevant (appropriate) thermal transport models.

Hence, some simplifications need to be made to solve this global problem. Since for device simulation purposes, we are only focused on calculating the *I–V* characteristics of a device accurately, the self-heating (which is a by-product of the current flowing through the device) may be treated approximately but still more accurately than the local heat conduction model.

Starting from the principle of energy conservation, Majumdar and coworkers [14, 15] derived separate energy balance equations for the optical phonon and the acoustic phonon bath. Under the application of electric fields greater than 10 kV/cm, electrons tend to lose energy primarily to optical phonons; optical phonons decay further to acoustic phonons. The energy conservation equations for optical and acoustic phonons are

$$\frac{\partial W_{\text{LO}}}{\partial t} = \left(\frac{\partial W_{\text{e}}}{\partial t}\right)_{\text{coll}} - \left(\frac{\partial W_{\text{LO}}}{\partial t}\right)_{\text{coll}} \tag{1.5}$$

and

$$\frac{\partial W_{\text{A}}}{\partial t} = \nabla(\kappa_{\text{A}} \nabla T_{\text{A}}) + \left(\frac{\partial W_{\text{LO}}}{\partial t}\right)_{\text{coll}} + \left(\frac{\partial W_{\text{e}}}{\partial t}\right)_{\text{coll}}, \tag{1.6}$$

where $W_e$, $W_{\text{LO}}$, and $W_{\text{A}}$ are electron, optical phonon, and acoustic phonon energy densities, respectively. Now

$$dW_{\text{LO}} = C_{\text{LO}} dT_{\text{LO}} \tag{1.7a}$$

and

$$dW_{\text{A}} = C_{\text{A}} dT_{\text{A}}, \tag{1.7b}$$

where $C_{\text{LO}}$ (specific heat capacity for optical phonons) can be estimated using the Einstein model, while $C_{\text{A}}$ (specific heat capacity for acoustic phonons) from the Debye model. Next, the collision terms are expressed using RTA:

$$\left(\frac{\partial W_{\text{e}}}{\partial t}\right)_{\text{coll}} = n \cdot \frac{\frac{3}{2} k_{\text{B}} T_{\text{e}} + \frac{1}{2} m^* v_{\text{d}}^2 - \frac{3}{2} k_{\text{B}} T_{\text{ph}}}{\tau_{\text{e-ph}}} \tag{1.8}$$

and

$$\left(\frac{\partial W_{\text{LO}}}{\partial t}\right)_{\text{coll}} = C_{\text{LO}} \frac{T_{\text{LO}} - T_{\text{A}}}{\tau_{\text{LO-A}}}, \tag{1.9}$$

where $T_{\text{e}}$ is the electron temperature, $v_{\text{d}}$ is the electron drift velocity, and $T_{\text{ph}}$ can be the optical or acoustic phonon temperature, depending on whichever kind of phonons the electrons interact with. Combining Eqs. 1.5–1.9, one arrives at

$$C_{\text{LO}} \frac{\partial T_{\text{LO}}}{\partial t} = \frac{3}{2} n k_{\text{B}} \left(\frac{T_{\text{e}} - T_{\text{LO}}}{\tau_{\text{e-LO}}}\right) + \frac{n m^* v_{\text{d}}^2}{2\tau_{\text{e-LO}}} - C_{\text{LO}} \left(\frac{T_{\text{LO}} - T_{\text{A}}}{\tau_{\text{LO-A}}}\right) \tag{1.10}$$

and

$$C_{\text{A}} \frac{\partial T_{\text{A}}}{\partial t} = \nabla(\kappa_{\text{A}} \nabla T_{\text{A}}) + C_{\text{LO}} \left(\frac{T_{\text{LO}} - T_{\text{A}}}{\tau_{\text{LO-A}}}\right) + \frac{3}{2} \frac{n k_{\text{B}}}{\tau_{\text{e-A}}} (T_{\text{e}} - T_{\text{A}}) + \frac{1}{2} \frac{n m^* v_{\text{d}}^2}{\tau_{\text{e-A}}}. \tag{1.11}$$

The first two terms on the right-hand side (RHS) of Eq. 1.10 represent the energy gained from the electrons, where $n$ is the electron density, $T_e$ is the electron temperature, $v_d$ is the drift velocity, and $T_{LO}$ is the optical phonon temperature, while the last term is the energy lost to the acoustic phonons. The same term also appears as a gain term on the RHS of Eq. 1.11. The first term on the RHS of Eq. 1.11 describes heat diffusion.

For the case where the electric fields are less than 10 kV/cm, electrons lose energy directly to the acoustic phonons and in that case, the energy balance equations can be expressed as

$$C_A \frac{\partial T_A}{\partial t} = \nabla(\kappa_A \nabla T_A) + \left(\frac{\partial W_e}{\partial t}\right)_{coll} \tag{1.12}$$

and

$$C_A \frac{\partial T_A}{\partial t} = \nabla(\kappa_A \nabla T_A) - \frac{3}{2}\frac{nk_B}{\tau_{e-A}} T_A + n \cdot \frac{\frac{3}{2}k_B T_e + \frac{1}{2}m^* v_d{}^2}{\tau_{e-A}}. \tag{1.13}$$

Under the assumption of very low electric fields, the electron temperature and the acoustic phonon temperatures equal the lattice temperature. Hence, the second and third terms in Eq. 1.13 cancel. Using the low field conductivity and the mobility expressions, the heat source term in Eq. 1.4 reduces to the last term of Eq. 1.13, as given here:

$$q_{gen} = \mathbf{J} \cdot \mathbf{E} = \sigma E^2 = \frac{\sigma v_d^2}{\mu^2} = \frac{nm^* v_d^2}{\tau} \tag{1.14}$$

It is assumed here that for low doping concentrations, the relaxation time, $\tau$, in Eq. 1.14 is the acoustic phonon relaxation time. The reason for this assumption is that the acoustic phonon scattering process, being isotropic in nature, is mostly effective in randomizing the carrier momentum while the carrier energy is very low (under the application of low electric fields). The local Joule heating approximation given by the result in Eq. 1.14 is only valid for low fields, which is not the case in nanoscale devices.

While considering the electron-lattice coupling in Eqs. 1.10 and 1.11, energy transfer from energetic electrons to optical phonons is very efficient. However, optical phonons have a negligible group velocity and, therefore, do not participate much in the heat diffusion process effectively. Instead, they transfer their energy to acoustic phonons, which can diffuse heat much more effectively. The energy transfer between phonons is a relatively slow process compared to the electron-optical phonon transport, and therefore a thermal nonequilibrium condition may also exist between optical and acoustic phonons. Figure 1.2 shows the primary thermal energy transport path and the corresponding time constants.

In the current, state-of-the-art electrothermal simulator, details of which can be found in the next section, steady-state versions of Eqs. 1.10 and 1.11 for the optical and acoustic phonon temperatures are solved self-consistently with a Monte Carlo (MC) simulation tool for the electron Boltzmann transport equation. This tool has been used to study self-heating in different technology nodes of nanoscale FD-SOI devices and DG device structures. As has already been explained, in smaller devices, the nonstationary transport and velocity overshoot effects dominate the carrier transport, and the results from the simulation tool suggest a less degrading effect of self-heating on the on-current characteristics.

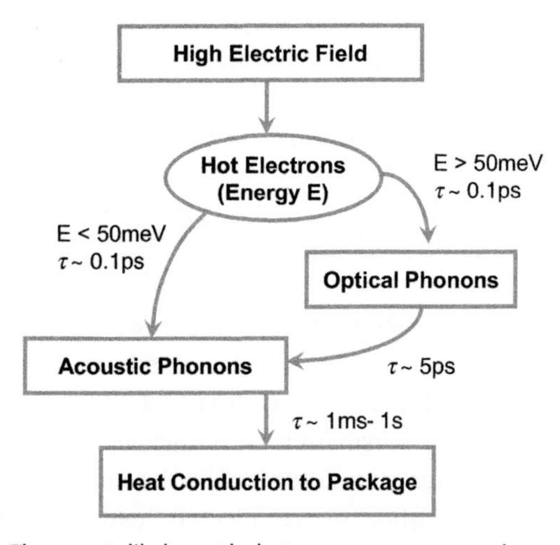


**Figure 1.2** The most likely path between energy carrying particles in a semiconductor device is shown, together with the corresponding scattering time constants.

### 1.2.2 Arizona State University Model Description

In the theoretical model described below, the standard ensemble Monte Carlo (EMC) code for the carrier Boltzmann transport equation solution has been modified as well. Figure 1.3 presents the flowchart of the modified EMC code.

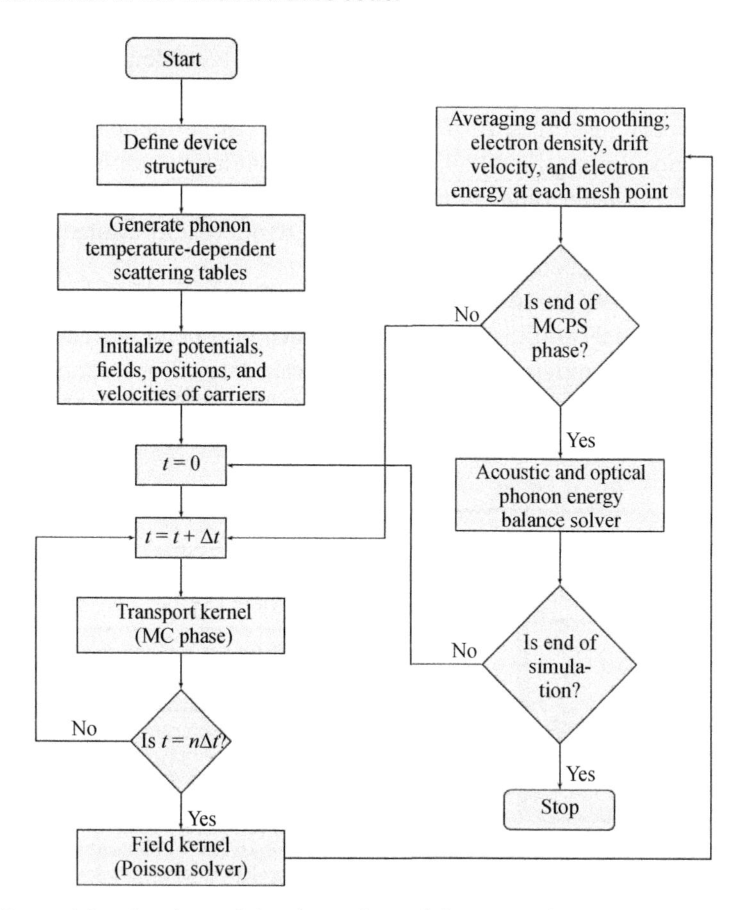


**Figure 1.3** Flowchart of the electrothermal device simulator coupling of the Monte Carlo solver for electrons and the energy balance equations for acoustic and optical phonons.

As there are variable lattice temperatures in the hot spot regions, the concept of regional scattering tables has been introduced. For each acoustic phonon temperature, one energy-dependent scattering table is created. These scattering tables involve additional steps in the MC phase (Fig. 1.4, right) since, to choose a scattering mechanism for a given electron energy randomly, it is necessary to know the corresponding scattering table. As depicted in Fig. 1.4 (right), the electron position on the grid needs to be known, so as to retrieve the acoustic and optical phonon temperatures at that grid point, and then the scattering table with coordinates ($T_L$,$T_{LO}$) is selected. Precalculation of the energy- and temperature-dependent scattering tables does not require much central processing unit (CPU) time or memory resources. Also, the scattering table formation is done only once during the initialization of the simulation and is carried out for a broader range of temperatures. For the temperatures, for which the appropriate scattering tables are not readily available, an interpolation scheme is employed.

After the MC phase is complete, a time averaging of carrier density, drift velocities, and thermal energy at each grid point is done and the electron temperature distribution is calculated. The exchange of variables between electron and phonon solvers is shown in the left panel of Fig. 1.4.

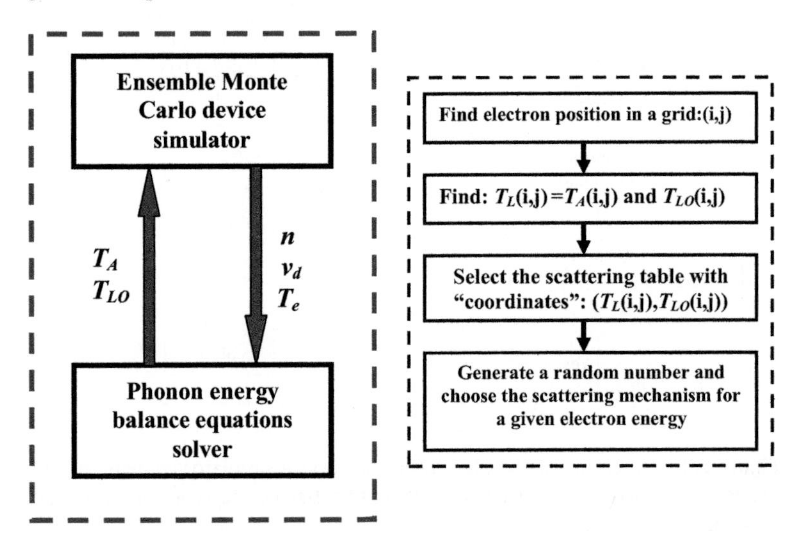


**Figure 1.4** (Left) Exchange of variables between the two kernels. (Right) Choice of the proper scattering table.

Note that in several of the studies, presented under simulation results (next section), we have considered SOI devices that consist of two distinct regions: silicon device layer and the buried oxide (BOX) layer. Since phonons have significantly smaller mean free paths within the oxide layer, the phonon energy balance equations for the acoustic and optical phonons are solved in the silicon layer to accurately model the heat transport. A simpler heat diffusion equation is used in the amorphous BOX. This is because the characteristic length is much smaller than the actual thickness of the film. The two computational regions are then coupled through interface conditions that account for differences in their respective material properties. It is necessary to calculate the flux of energy, passing through the interface for each time step, between the two material systems at each point along the interface.

To properly solve the phonon balance equations, the device should be attached to a heat sink somewhere along the boundary or finite heat conduction through the surface should be allowed for. In our code, a heat sink is modeled by a simple Dirichlet boundary condition (i.e., constant temperature). We use the substrate electrode and the bottom of the BOX as a heat sink.

## 1.3 Simulation Results

### 1.3.1 Self-Heating Effects in FD-SOI Devices

#### 1.3.1.1 Basic findings

The electrothermal device simulator described in Section 1.2 was initially used to study the effects of self-heating on the electrical characteristics in different generations of nanoscale FD-SOI devices (Fig. 1.5). Simulation results suggest that self-heating has a less degrading effect on the on-current in smaller devices in which the nonstationary transport and velocity overshoot effects dominate the carrier transport. Due to the quasi-ballistic nature of the carrier transport, there is less scattering in the active channel region; therefore, there is a smaller probability of the transfer of energy to the lattice. In these nanoscale structures, most of the excess carrier energy is dissipated at the drain contact. This trend is clearly seen from the snapshots of the lattice temperature profiles presented in Fig. 1.6. From these results one can observe that (a) the temperature in the channel is increasing with the increase of the channel length and (b) the maximum lattice temperature region (hot spot) is in the drain and it shifts toward the channel for larger devices. This is more pronounced for a higher temperature on the gate contact.

| $L$ (nm) | $t_{ox}$ (nm) | $t_{Si}$ (nm) | $t_{box}$ (nm) | $N_{ch}$ (cm$^{-3}$) | $V_{GS}=V_{DS}$ (V) | $I_D$ (mA/um) |
|---|---|---|---|---|---|---|
| 25 | 2 | 10 | 50 | $1\times10^{18}$ | 1.2 | 1.82 |
| 45 | 2 | 18 | 60 | $1\times10^{18}$ | 1.2 | 1.41 |
| 60 | 2 | 24 | 80 | $1\times10^{18}$ | 1.2 | 1.14 |
| 80 | 2 | 32 | 100 | $1\times10^{17}$ | 1.5 | 1.78 |
| 90 | 2 | 36 | 120 | $1\times10^{17}$ | 1.5 | 1.67 |
| 100 | 2 | 40 | 140 | $1\times10^{17}$ | 1.5 | 1.57 |
| 120 | 3 | 48 | 160 | $1\times10^{17}$ | 1.8 | 1.37 |
| 140 | 3 | 56 | 180 | $1\times10^{17}$ | 1.8 | 1.23 |
| 180 | 3 | 72 | 200 | $1\times10^{17}$ | 1.8 | 1.0 3 |

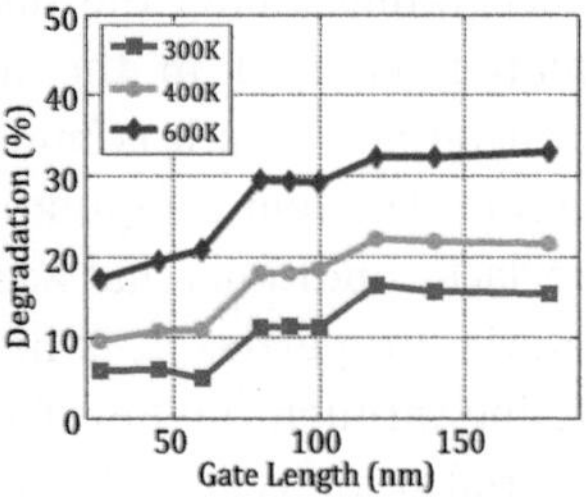


**Figure 1.5** (Left) Parameters for various simulated FD-SOI device technology nodes (constant field scaling http://public.itrs.net/). $L$, gate length; $t_{ox}$, gate oxide thickness; $t_{Si}$, active Si layer thickness; $t_{box}$, BOX thickness; $N_{ch}$, channel doping concentration; $I_D$, isothermal current value (300 K). (Right) Current degradation versus technology generation ranging from 25 nm to 180 nm channel length FD-SOI devices. The isothermal boundary condition of 300 K is set on the bottom of the BOX. The parameter is the temperature on the gate contact. Neumann boundary conditions are applied at the vertical sides and at the source/drain contacts.

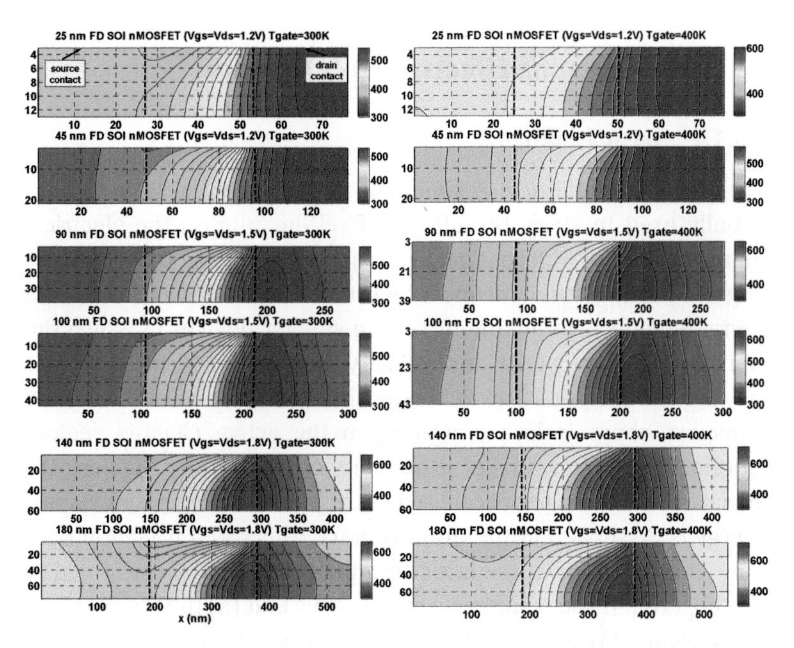


**Figure 1.6** Lattice temperature profiles for SOI devices of different technology generations with gate lengths ranging from 25 nm to 180 nm. Neumann boundary conditions are applied on the source, drain, and artificial side boundaries. The temperature on the gate contact is 300 K (left) and 400 K (right).

#### 1.3.1.2 Thermal boundary conditions and proper choice of the device simulation domain

In obtaining the results from Figs. 1.4–1.6, we have assumed Neumann boundary conditions (no heat flow) at the source/drain contacts and the artificial (side) boundaries and Dirichlet boundary conditions (heat sink) at the gate electrode.

Next we extend our analysis to devices from different technology generations and different boundary conditions at the source/drain, the gate electrode, and the artificial boundaries to examine trends in device behavior with scaling. The simulation results can be summarized as follows:

- With regard to the role of thermal boundary conditions on the gate and side electrodes, we consider separately analog and digital circuits. In analog devices, neighboring devices are typically on and if the gate contacts are also biased then there is no heat flow through the gate contact and the side boundaries. In this case it is appropriate to use Neumann boundary conditions on the side (artificial) boundaries and Neumann boundary conditions on the gate electrode. In the case of digital devices where the operation mode of the transistors changes in certain intervals from on to off to on, Dirichlet boundary conditions at the gate and the side boundaries are appropriate boundary conditions. This corresponds to the best-case scenario of heat removal from the device active region. Simulation results for the current degradation for different technology devices when Dirichlet/Neumann boundary conditions are applied to the gate and the side boundaries are shown in Fig. 1.7. One can observe that when the gate electrode is used as an ideal heat sink ($T_{\text{gate}}$ = 300 K), there is no difference in the current degradation due to a self-heating with thermal Neumann or Dirichlet boundary conditions applied at the vertical sides. But, when heat transfer is not allowed through the gate electrode (Neumann boundary condition), the use of different boundary conditions at the vertical sides has an impact on the current degradation only for smaller devices. The use of Dirichlet boundary conditions at both vertical sides and the gate electrode is more adequate for. The worst-case scenario for self-heating is when Neumann boundary conditions are imposed on all outer boundaries, except on the bottom of the BOX.

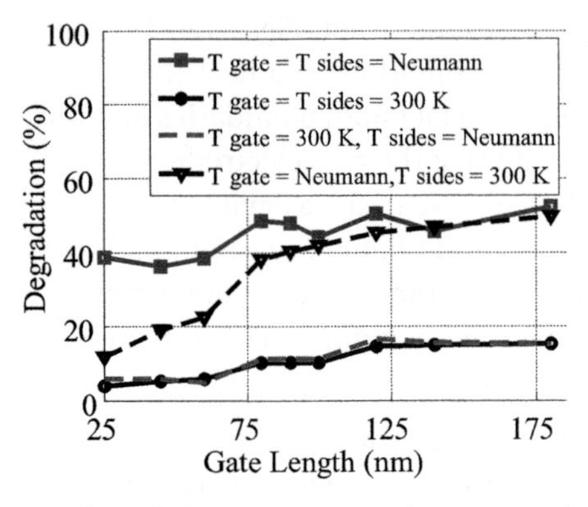


**Figure 1.7** Current degradation versus technology generation ranging from 25 nm to 180 nm channel length FD-SOI devices for different combinations of Dirichlet and Neumann thermal conditions at the gate electrode and vertical sides.

- To further investigate whether Neumann boundary conditions are the appropriate choice of boundary conditions for the source and drain electrode and also to examine the role of the finite contact size, we extend the length of the metal gate and source and drain electrodes and assume the best-case scenario; namely, we neglect the Joule heating. The simulation results (see Ref. [16]) show that the gate electrode does not need to be modeled unless the gate is near a hot spot that would prevent thermal conduction. However, source and drain extensions play a significant role in the heat transfer into the interconnects. Therefore, if extended source and drain electrodes are simulated without the Joule heating term, that would correspond to the best heat conduction scenario. If one assumes Neumann boundary conditions at the source and drain electrodes, then that would be the worst-case scenario with respect to the modeling of the contacts. A real device operates somewhere between these two limits. The worst-case scenario would be Neumann boundary conditions everywhere except at the substrate electrode, but such operating conditions are rather rare. Thus, everything depends on whether the device is near a hot spot or not. If it is, then Neumann boundary conditions are appropriate and if it is not then some sort of leaky Dirichlet boundary conditions at the source and drain electrode are appropriate.

## 1.3.2 Can We Reduce Self-Heating?

### 1.3.2.1 Single-gate versus dual-gate FD-SOI devices

To get better device performance and eliminate threshold voltage fluctuations and other imperfections that arise during the fabrication of nanoscale devices, researchers in the last 15 years have focused on alternative transistor designs such as DG devices, trigate (FinFET) devices, and multigate devices. The first proposal for DG devices was vertical structures (see Fig. 1.8 and Ref. [17]), whereas later proposals involve in-plane devices that are easily integrated in the complementary metal-oxide-semiconductor (CMOS) process. Since at the time when this research work was performed we only had a 2D electrothermal particle-based device simulator, we focus on the comparison of performance degradation due to self-heating effects in single-gate (SG) and vertical DG devices (both electrical flow and heat flow occur in a 2D plane).

The on-current degradation for different boundary conditions on the gate electrode in a single-gate fully depleted silicon-on-insulator (SG FD-SOI) device (Fig. 1.8, left) and the corresponding current degradation for a DG device (Fig. 1.8, right) and different boundary conditions on the temperature of both the top and the bottom gates are shown in Table 1.1. It is found that the DG device has almost the same current degradation as the SG device, even though it carries about 1.5–1.8 times more current. However, the dual-gate device has a higher lattice temperature in the hot spot region and there is a larger bottleneck between acoustic and optical phonons in the DG device structure when compared to the SG FD-SOI device structure. This can be easily explained by the fact that the DG structure holds more carriers and since the decay process from optical to acoustic phonon is not fast enough, heating has a higher influence on the carrier drift velocity and, consequently, the on-state current in DG devices. In fact, degradation is observed in the average carrier velocity in the DG devices when compared to a SG FD-SOI device structure (Fig. 1.9).

**Table 1.1** Current degradation due to self-heating for single-gate (SG) and dual-gate (DG) structures given in Fig. 1.8, for $V_{GS}$ = 1.2 V and $V_{DS}$ = 1.2 V. Temperature boundary conditions on the DG device gate electrodes are set to 300 K

| | 25 nm SG FD-SOI | | 25 nm DG FD-SOI | |
|---|---|---|---|---|
| **Type of simulation** | **Current (mA/μm)** | **Current decrease (%)** | **Current (mA/μm)** | **Current decrease (%)** |
| Isothermal | 1.94 | NA | 3.1 | NA |
| Thermal (300 K) | 1.76 | 9.18% | 2.79 | 9.13% |
| Thermal (400 K) | 166 | 14.35% | 2.63 | 14.37% |
| Thermal (600 K) | 1.50 | 22.82% | 2.32 | 24.54% |

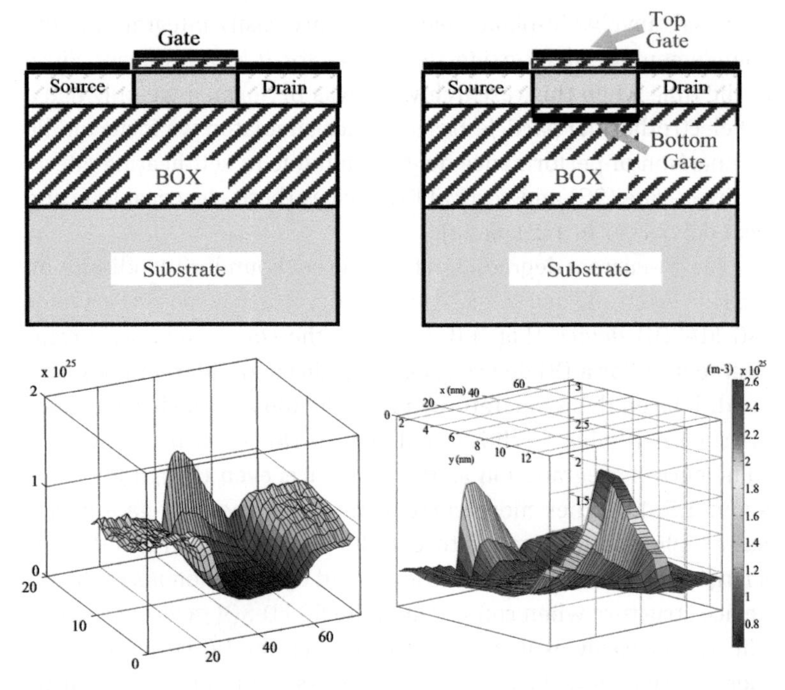


**Figure 1.8** (Top) Schematic of the single-gate (left) FET and the dual-gate (right) FET structure being modeled, with (bottom) the corresponding electron densities. Note the two separate channels for the dual-gate structure. The doping/geometrical dimensions of these two structures are $N_D = 1 \times 10^{25}$ m$^{-3}$ (source and drain region), $N_A = 1 \times 10^{17}$ m$^{-3}$ (channel region), $L_{gate}$ = 25 nm (channel length), $t_{ox}$ = 2 nm (gate oxide thickness), $t_{si}$ = 12 nm (Si layer thickness), and $t_{BOX}$ = 50 nm (BOX thickness).

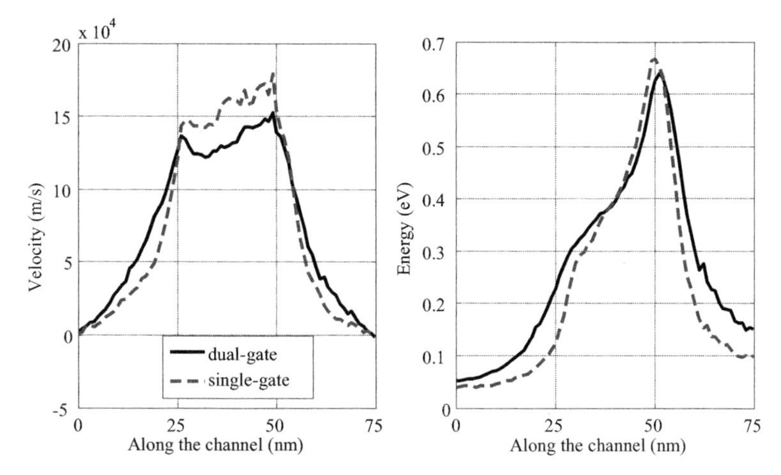


**Figure 1.9** Average electron velocity (left) and average electron energy (right) in 25 nm channel length single-gate and dual-gate SOI devices.

### 1.3.2.2 FD-SOI devices with diamond and AlN BOX

Improving device performance from an electrical as well as thermal perspective does not depend only on the use of multigate device structures. There also exists an alternative approach to achieving the same goal, that is, reduction in the thermal resistance of the BOX. Studies have found many suitable materials that can serve as BOX layers. Some of the frequently studied materials for the BOX layer are diamond, AlN, SiC, etc.

SOD is a substrate engineered specifically to address the major challenges of silicon-based ultra-large-scale integrated (ULSI) technology. Particularly, it provides enhanced thermal management and charge confinement. The SOD concept is achieved by joining a thin, single-crystalline Si device layer to a highly oriented diamond (HOD) layer that serves not only as an electrical insulator but as a heat spreader and supporting substrate as well. Hence SOD can be used as an alternative material in SOI devices, where the thermally insulating $SiO_2$ is replaced by highly thermally conductive diamond.

An alternative approach is to replace the buried silicon dioxide by another insulator that has higher thermal conductivity. One of the interesting candidates for such novel buried insulators is aluminum nitride (AlN), which has thermal conductivity that is about 100 times higher than that of $SiO_2$ (136 W/m-K versus 1.4 W/m-K) and roughly equal to that of bulk silicon itself (145 W/m-K). Furthermore, AlN has excellent thermal stability, high electrical resistance, and a coefficient of thermal expansion that is close to that of silicon. Thus, the use of AlN as the buried insulator in silicon-on-aluminum nitride (SOAlN) as an alternative to SOD may help mitigate the self-heating penalty of the SOI materials, hence enabling more general and higher temperature applications.

Simulation results for 25 nm channel length fully depleted silicon-on-diamond (FD-SOD) and fully depleted silicon-on-aluminum-nitride (FD-SOAlN) devices show (i) lattice heating plays a minor role in current degradation compared to FD-SOI devices (see Fig. 1.10) and (ii) the spread of temperature on the bottom of the wafer is more uniform. This means that if adopted, hot spots are less likely to occur in these two device technologies.

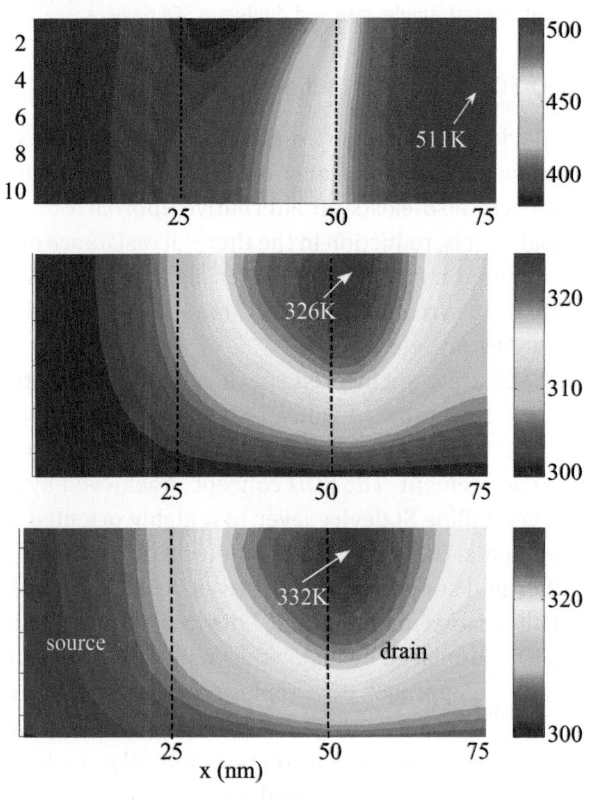


**Figure 1.10** Lattice temeperature profile in the active silicon layer for 25 nm channel length SOI (top), SOD (middle), and SOAlN MOSFET structure.

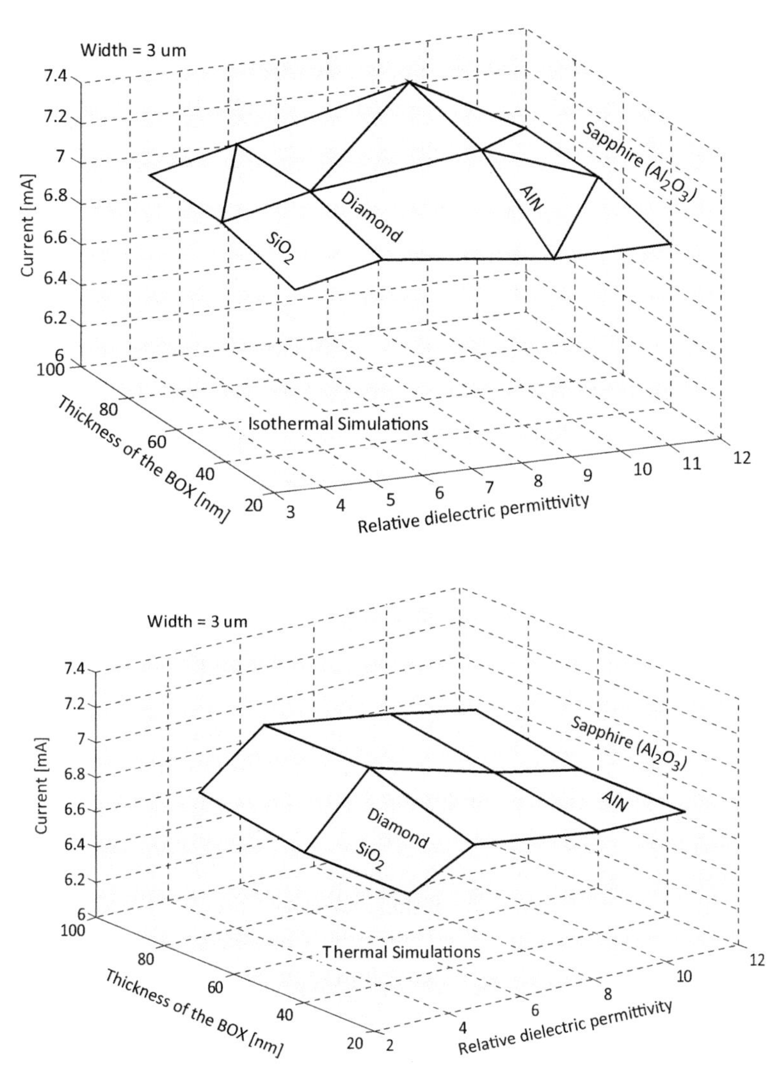


**Figure 1.11** (Top) Isothermal simulations and (bottom) thermal simulations. Optimization is performed by varying the thickness of the box.

Furthermore, a series of experiments were designed for the four possible BOX materials. The BOX thickness was varied in order to determine which material had the best performance. The values for the current for a device with a 3 μm width are summarized in Fig. 1.11. The simulations with 10 Gummel cycles represent thermal situation whereas 1 Gummel cycle represents the isothermal case. When only the isothermal situation is considered, AlN is the material of choice for the BOX. However, when thermal effects are accounted for, then diamond wins as a material of choice.

### 1.3.3 Multiscale Modeling: Modeling of Circuits (CS and CD Configuration)

The International Technology Roadmap for Semiconductors (ITRS) suggests that as devices are scaled to smaller dimensions, the current density in the interconnects would increase [18, 19]. Hence it is important to account for heating effects not only within the device itself but also within the contacts and interconnects when considering reliability of a system. To do exactly that, a novel multiscale simulation approach that combines circuit-level simulations with device-level simulations has been proposed. The approach compares simulation results with experimental measurements in an attempt to uncover the temperature profile due to self-heating effects. The proposed method couples circuit-level simulation performed using Silvaco Atlas [20] with an electrothermal MC device simulator [21]. The Giga3D Silvaco Atlas module simulates the thermal transport characteristics at the interconnect level. This module provides temperature boundary conditions for the device-level simulation. Then the device-level simulator solves for self-heating throughout the device. The coupled system is shown in Fig. 1.12. It is important to note that in the device-level thermal solver, the 2D/3D Poisson's equation is solved self-consistently with an MC transport kernel coupled to a 2D/3D energy balance equations solver for the acoustic (lattice) and optical phonon baths [22], as described in Section 1.2. This is by itself a new multiscale approach that is very different from the commonly used Joule heating model used in commercial device simulators. Such simulations give rise to more pronounced hot spots because they accurately represent the optical-to-acoustic-phonon bottleneck [23].

The device-level temperature measurement technique used in this work is based on the temperature dependence of the subthreshold slope of a planar MOSFET. The underlying idea is that variations in the subthreshold slope can be used to determine the temperature within the hot spot. Two devices in either common-source or common-drain configuration are considered. One device functions as a heater, while the other functions as a sensor. The subthreshold slope at the sensor side varies in direct response to the temperature at the heater device side. A sample of biasing conditions, together with the schematics for both configurations, is given in Fig. 1.13 (left). Figure 1.13 (right) shows the mask image that indicates that both FETs are located in a common active area and are separated by only one poly pitch.

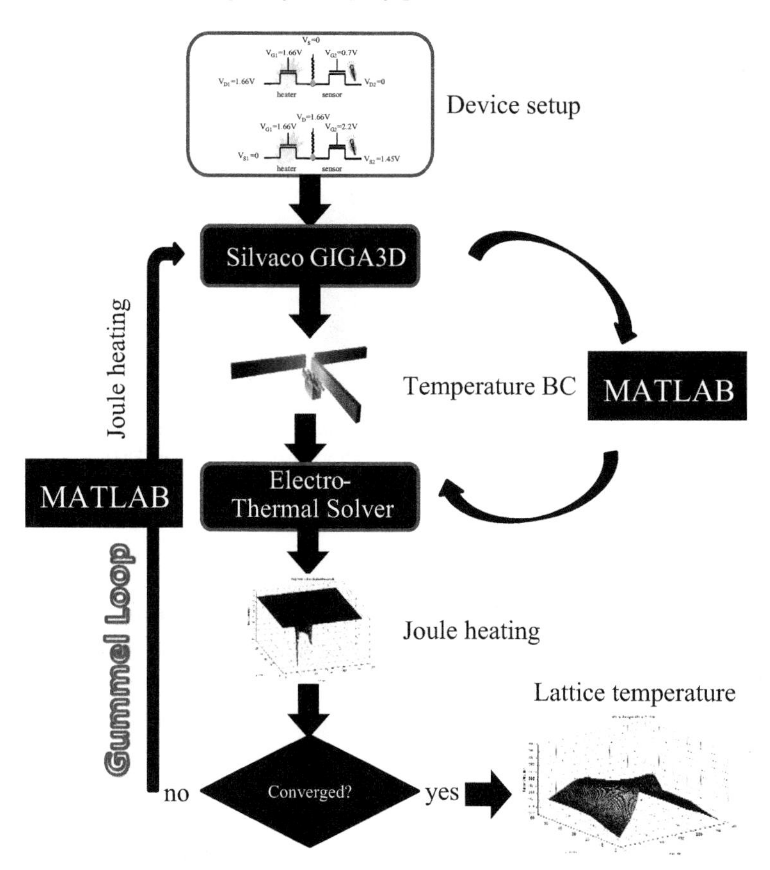


**Figure 1.12** Implemented multiscale modeling scheme.

The experimental procedure to estimate the hot spot temperature was proposed by IMEC group [24]. First, as schematically illustrated in Fig. 1.14, the increase in temperature ($\Delta T$) induced by an nFET (i.e., the heater from Fig. 1.13) is extracted by making use of an nFET sensor that is located near this nFET or the device under test (DUT). This is done by using temperature dependent characteristics of the sensor. As can be seen, the sensor is connected to a common-source configuration with the heater (Fig. 1.13, top-left and right). This configuration allows for the closest possible in-silicon sensor since the two devices are separated only by one gate pitch. Also, both these devices share the same active area, which is surrounded by shallow trench isolation (Fig. 1.13, right). Subsequently, a subthreshold swing (SS) in the sensor is extracted using a modified Enz–Krummenacher–Vittoz (EKV) model, as illustrated in Fig. 1.14 [25].

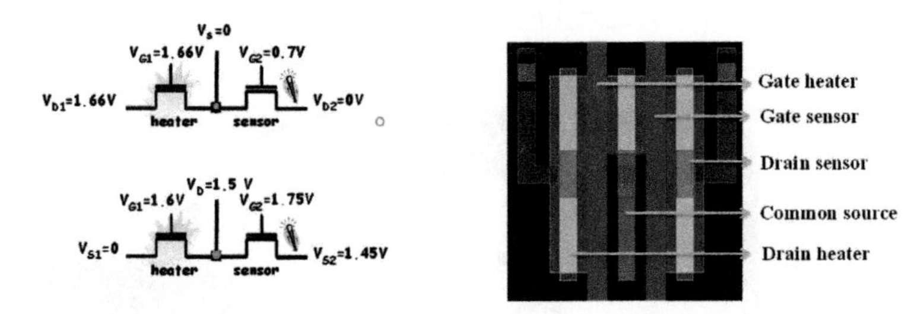


**Figure 1.13** (Left) Two possible measurement configurations are common source (CS) (top) and common drain (CD) (bottom). The heater (DUT = device under test) operates in saturation, while the sensor operates in the subthreshold region. (Right) Mask image indicating that both FETs are located in a common active area, spaced by only one poly pitch.

The results from the experimental data are used as a reference for the thermal simulations. The coupled solver uses the MC method and the energy balance and Poisson's equation to simulate the heating at the device level. Joule heating, defined as a product of the current density and the electric field, is extracted from these device-level simulations and used in the circuit-level simulation. The Joule heating term is used as an input for the interconnect-level solver (Giga3D module within the Silvaco Atlas framework). This Silvaco model provides the temperature boundary conditions at the device-level within the global electrothermal device simulator. To integrate and interface these two separate modules, MATLAB is used (shown in Fig. 1.12).

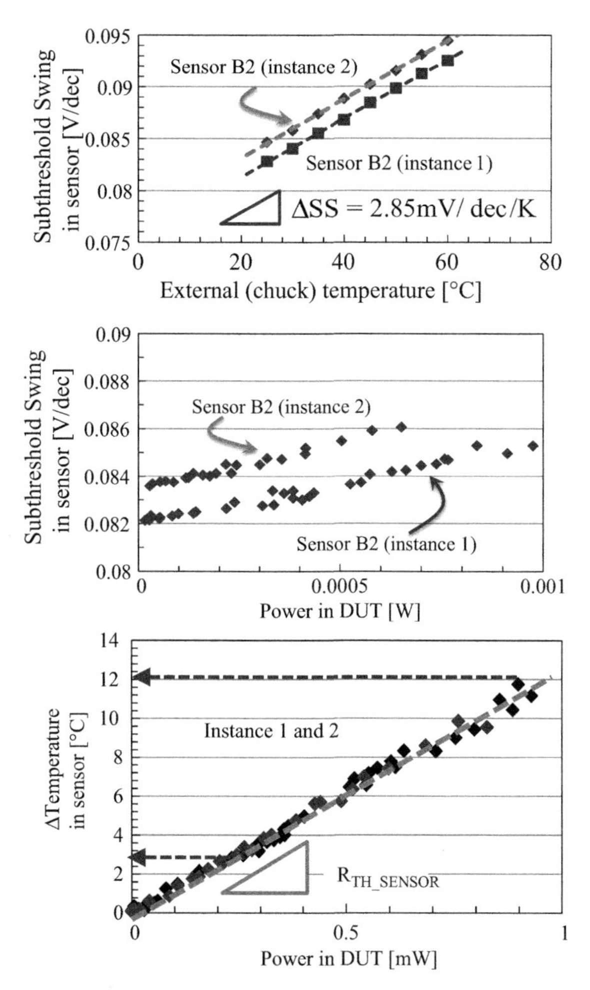


**Figure 1.14** Results for two instances of a device from Fig. 1.13. (Top) The subthreshold swing (SS) varies linearly with the externally applied chuck temperature. (Middle) When drawing a large current through the heater, the SS of the sensor varies linearly. (Bottom) Using the initial SS as a reference, the extracted $\Delta T$ in the sensor gives consistent results for both instances.

The experimental transfer characteristic curves for drain voltages $V_{DS}$ = 1, 1.5, and 2 V at the sensor are shown in Fig. 1.15 (left). From the extracted subthreshold slopes, using the EKV method [23, 26], the corresponding average sensor temperatures for different bias conditions ($V_{DS}$ and $V_{GS}$) are shown in Fig. 1.15 (right). The goal of the multiscale simulations is twofold [27]: (i) Given the structure and the bias conditions in the heater-sensor configuration, the sensor temperature variation from Fig. 1.15 is reproduced, and (ii) once the sensor temperature match is achieved, extrapolation of the peak heater (DUT) temperature is performed. Hence the temperature of the hot spot is uncovered in an indirect way.

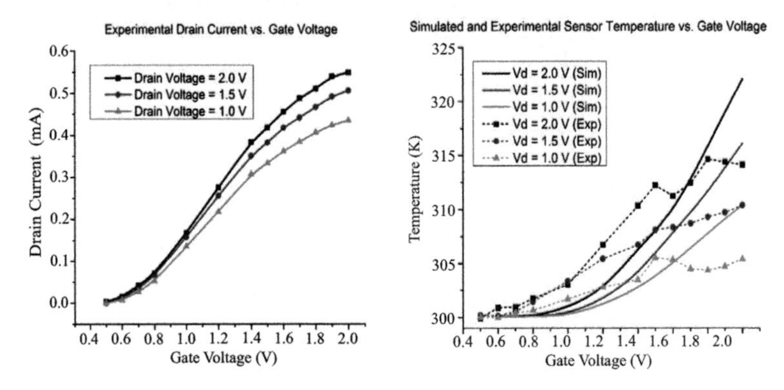


**Figure 1.15** (Left) Measured transfer characteristics of the sensor. The parameter is the drain voltage $V_{DS}$ = 1, 1.5, and 2 V. (Right) Extrapolated and simulated average sensor temperature for different combinations of drain and gate voltage. The EKV method [23, 24] is used for the extrapolation of the sensor temperature from the subthreshold characteristics from Fig. 1.14.

To account for the heating at the source, gate, and drain interconnects as shown in Fig. 1.12, first the complete circuit (device + interconnects) domain is solved using the Giga3D module of Silvaco Atlas. The temperature at the interconnects as well as the temperature in the device (along one cross section) are shown in Fig. 1.16. At the boundaries of the rectangular cross section from the bottom panel in Fig. 1.16, the lattice temperature is registered and used in the thermal particle-based device simulator, which then provides (i) the actual temperature of the hot spot and (ii) the Joule heating terms that are used back in the Giga3D module for the next Gummel iteration (outer loop) of the model. It should be noted that solving the energy balance equations for acoustic and optical phonons self-consistently with the Boltzmann transport equation for the electrons (that is solved using the MC method) is, by itself, a multiscale problem [28]. Hence, in the implemented scheme we have three levels of abstraction. The convergence of the global Gummel loop is shown in Fig. 1.17.

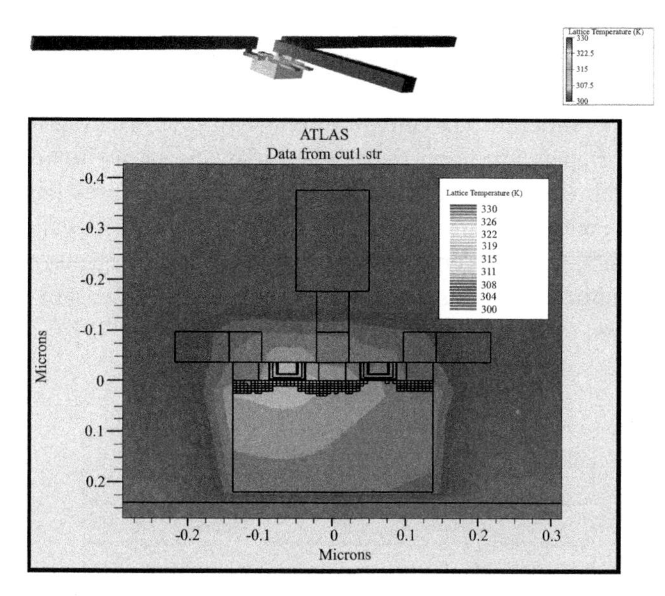


**Figure 1.16** (Top) Giga3D modeling of the device + interconnects. (Bottom) Extracted lattice temperature boundary conditions from Giga3D simulations. Bias conditions for the common-drain configuration are $V_S$ (DUT) = 0, $V$ (common) = 1.5 V, $V_G$ (DUT) = 1.6 V, $V_G$ (sensor) = 1.75 V, and $V_S$ (sensor) = 1.45 V, as shown in Fig. 1.13 (bottom-left).

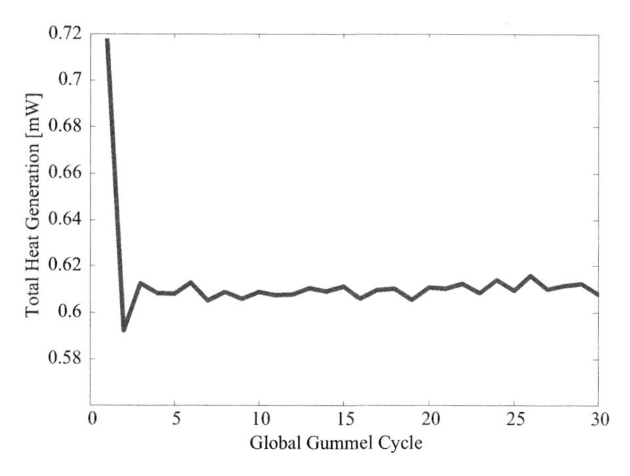


**Figure 1.17** Convergence of the global Gummel cycle (loop).

The global loop converges within 5–10 Gummel cycles. In Figs. 1.18 and 1.19, the lattice (acoustic phonon) temperature and the optical phonon temperatures for the heater-sensor combination in the common-drain configuration (Fig. 1.13, bottom-left) are shown. Depending upon the applied bias, the lattice temperature profile obtained from Silvaco Atlas leads to an underestimation of the hot spot temperature by about 10–20 K. Figure 1.19 shows the bottleneck in the energy transport due to the low group velocity of the optical phonons. One can see a more localized hot spot as compared to the acoustic (lattice) temperature case [29].

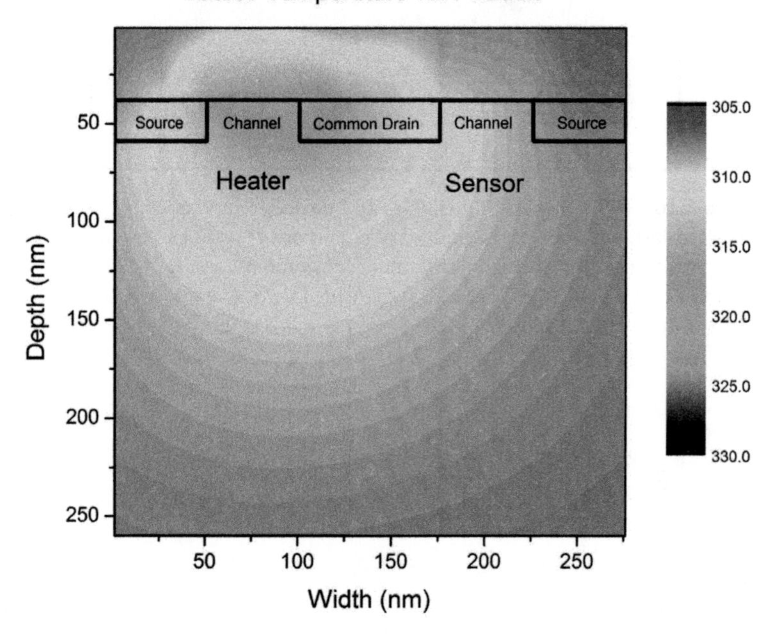


**Figure 1.18** Lattice temperatures in the heater (DUT)-sensor configuration.

## 1.4 Conclusions and Future Directions of Research

In summary, we have presented a way of modeling self-heating in nanoscale devices using our electrothermal device simulator,

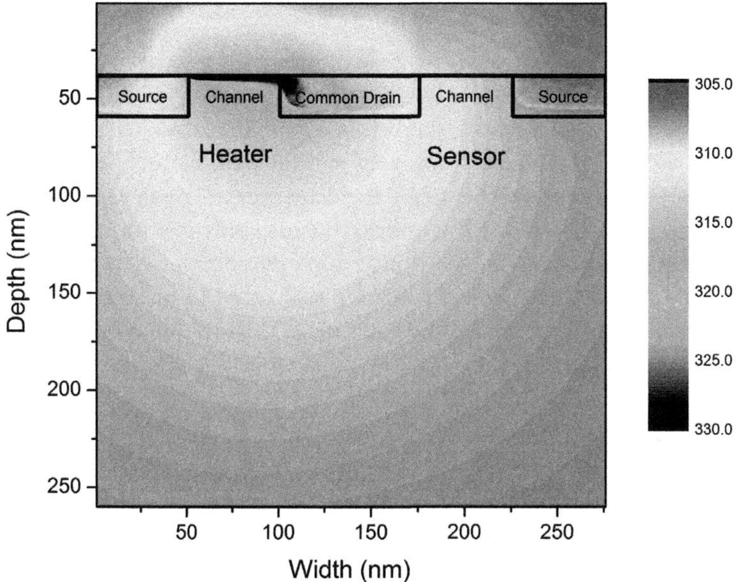


**Figure 1.19** Optical phonon temperatures in the heater (DUT)-sensor configuration.

which by itself is a multiscale solver. We have applied the solver on understanding self-heating effects in FD-SOI devices, DG SOI devices, and SOI devices with different BOX layers. It is important to note that the electron problem is limited to length scales that are on the order of several tens of nanometers. However, the phonon mean free path in bulk Si is on the order of 300 nm at room temperature, which is a much larger length scale. Hence, simulations of lattice heating require a much larger simulation domain. In addition to this, as we scale semiconductor devices into smaller dimensions, the role of interconnects' self-heating has to be accounted for as well. For that purpose we have introduced another level of hierarchy in the multiscale modeling approach. We model with Giga3D (Silvaco Atlas module) the role of interconnects and the role of the larger simulation domain, and we extract the temperature boundary conditions on a smaller domain in which we use our electrothermal device simulator, which, in turn, provides Joule heating terms to Giga3D. The whole Gummel iteration loop is repeated until a self-consistent solution at multiple levels of approximation in our theoretical models is reached. This is a first simulation of this type in the world.

In the future, we propose to extend or electrothermal device simulator as depicted in Fig. 1.20. This will involve, first, use of the thermal conductivity data that are calculated using large-scale atomic/molecular massively parallel simulator (LAMMPS) simulation software in the original electrothermal device simulator. Second, we will replace the energy balance solver with a phonon Boltzmann transport solver that is already completed within our group. LAMMPS simulation software will be used to calculate the proper phonon band structure that is input to the phonon Boltzmann transport equation solver.

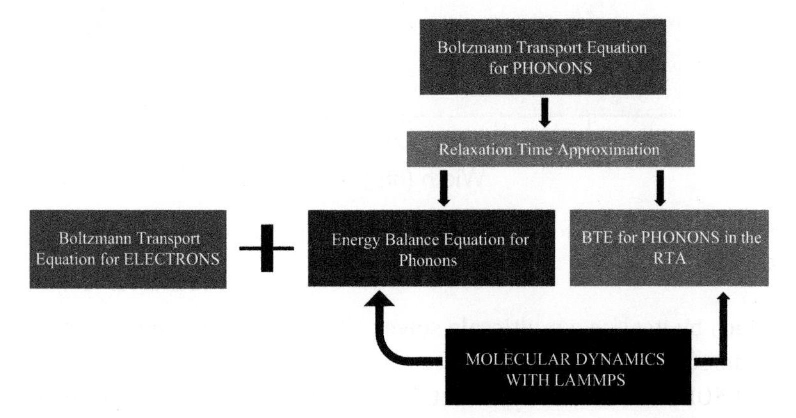


**Figure 1.20** Proposal for extending the electrothermal solver at the Arizona State University.

## References

1. Fayneh, E., Yuffe, M., Knoll, E., Zelikson, M., Abozaed, M., Talker, Y., Shmuely, Z., and Rahme, S. A. (2016). 14nm 6th-generation core processor SOC with low power consumption and improved performance, *IEEE International Solid-State Circuits Conference*, pp. 72–74.

2. Song, T., Rim, W., Park, S., Kim, Y., Jung, J., Yang, G., Baek, S., Choi, J., Kwon, B., Lee, Y., Kim, S., Kim, G., Won, H. S., Ku, J. H., Paak, S. S., Jung, E., Park, S. S., and Kim, K. A. (2016). 10nm finfet 128mb sram with assist adjustment system for power, performance, and area optimization, *IEEE International Solid-State Circuits Conference*, pp. 306–309.

3. Mittal, A., and Mazumder, S. (2010). Monte carlo study of phonon heat conduction in silicon thin films including contributions of optical phonons, *J. Heat Transfer*, **132**, p. 052402.
4. Jang, D., Bury, E., Ritzenthaler, R., Bardon, M. G., Chiarella, T., Miyaguchi, K., Raghavan, P., Mocuta, A., Groeseneken, G., Mercha, A., Verkest, D., and Thean, A. (2015). Self-heating on bulk finfet from 14nm down to 7nm node, *IEEE International Electron Devices Meeting (IEDM)*, pp. 11.6.1–11.6.4.
5. *International Solid-State Circuits Conference Trends*, http://http://isscc. org/trends/ (2015).
6. Maldovan, M. (2013). Sound and heat revolutions in phononics, *Nature*, **503**, pp. 209–217.
7. Kittel, C., and Kroemer, H. *Thermal Physics*, Freeman, New York, 1980.
8. Johnson, R. G., Snowden, C. M., and Pollard, R. D. (1997). A physics-based lectro-thermal model for microwav and millimetre wave HEMTs, *IEEE MTT-S Int. Microwave Symp. Dig.*, **3**, pp. 1485–1488.
9. Batty, W., Panks, A. J., Johnson, R. G., and Snowden, C. M. (2000). Electro-thermal modelling of monolithic and hybrid microwave and millimeterwave IC's, *VLSI Design*, **10**(4), pp. 355–389.
10. *Silvaco ATLAS Manual*, http://www.silvaco.org.
11. Datta, S. (1989). *Quantum Phenomena*, Addison-Wesley Modular Series of Solid State Devices, Vol. 8, Addison-Wesley, p. 89.
12. Liu, W. and Asheghia, M. (2004). Phonon–boundary scattering in ultrathin single-crystal silicon layers, *Appl. Phys. Lett.*, **84**, pp. 3819–3821.
13. Narumanchi, S. V. J., Murthy, J. Y., and Amon, C. H. (2004). Submicron heat transport model in silicon accounting for phonon dispersion and polarization, *Trans. ASME*, **126**, pp. 946–955.
14. Lai, J., and Majumdar, A. (1996). Concurent thermal and electrical modeling of submicrometer silicon devices, *J. Appl. Phys.*, **79**, pp. 7353–7361.
15. Majumdar, A., Fushinobu, K., and Hijikata, K. (1995). Effect of gate voltage on hot electron and hot phonon interaction and transport in a submicrometer transistor, *J. Appl. Phys.*, **77**, pp. 6686–6694.
16. Vasileska, D., Raleva, K., and Goodnick, S. M. (2009). Self-heating effects in nano-scale FD SOI devices: the role of the substrate, boundary conditions at various interfaces and the dielectric material type for the BOX, *IEEE Trans. Electron Devices*, **68**.

17. Hisamoto, D., Kaga, T., Kawamoto, Y., and Takeda, E. (1989).A fully depleted lean-channel transistor (DELTA): a novel vertical ultrathin SOI MOSFET, *IEDM Tech. Dig.*, p. 833.
18. www.itrs.gov
19. Sinha, S., and Goodson, K. E. (2005). Review: multiscale thermal modelling in nanoelectronics, *Int. J. Multiscale Comput. Eng.*, **3**(1), pp. 107–133.
20. www.silvaco.com
21. Raleva, K., Vasileska, D., Goodnick, S. M., and Nedjalkov, M. (2008) Modeling thermal effects in nanodevices, *IEEE Trans. Electron Devices*, **55**(6), pp. 1306–1316.
22. Vasileska, D., Goodnick, S. M., and Klimeck, G. (2010). *Computational Electronics: Semi-Classical and Quantum Transport Modeling*, CRC Press.
23. Majumdar, A., Fushinobu, K., and Hijikata, K. (1995). Effect of gate voltage on hot electron and hot phonon interaction and transport in a submicrometer transistor, *J. Appl. Phys.*, **77**, pp. 6686–6694.
24. Bury, E., Kaczer, B., Roussel, P. J., Ritzenthaler, R., Raleva, K., Vasileska, D., and Groeseneken, G. (2014). Experimental validation of self-heating simulations and projections for transistors in deeply scaled nodes, *Proc. IEEE, 2014 IEEE Int. Reliab. Phys. Symp.*, pp. XT.8.1–XT.8.6.
25. Enz, C. C., Krummenacher, F., and Vittoz, E. A. (1995). An analytical MOS transistor model valid in all regions of operation and dedicated to low-voltage and low-current applications, *Analog Integr. Circuits Signal Proc.*, **8**, pp. 83–114.
26. Enz, C. C., and Vittoz, E. A. *(2006).* Charge-Based MOS Transistor Modeling: The EKV Model for Low-Power and RF IC Design, Wiley.
27. Raleva, K., Bury, E., Kaczer, B., and Vasileska, D. (2014). Uncovering the temperature of the hotspot in nanoscale devices, *Proc. IEEE, 2014 Int. Workshop Comput. Electron. (IWCE)*, pp. 104–106.
28. Vasileska, D., Raleva, K., and Goodnick, S. M. (2010). Heating effects in nanoscale devices, in *Cutting Edge Nanotechnology*, edited by Dragica Vasileska, I-Tech Education, Kirchengasse, Austria.
29. Hossain, A., Vasileska, D., Raleva, K., and Goodnick, S. M. (2013). Interplay of self-heating and short-range coulomb interactions due to traps in a 10 nm channel length nanowire transistor, in *Nanoelectronic Device Applications Handbook*, edited by James Morris and Krzysztof Iniewski, CRC Press.

# Chapter 2

# Simulation of Charge and Thermal Transport

**Zlatan Aksamija**
*University of Massachusetts Amherst, Amherst, MA 01003, USA*
zlatana@engin.umass.edu

## 2.1 Introduction

To study charge and thermal transport at the nanoscale, we must consider not just the classical formulation of transport in the spatial domain but instead take together both real space and reciprocal, or momentum, space aspects of charge and thermal transport. To do this, we use the Boltzmann transport equation (BTE). This is a semiclassical formulation of transport, which is capable of including self-consistently the transport of both electrons and phonons, as well as externally applied electric fields, electron-phonon interaction, anharmonic phonon decay, and many other types of scattering to a desired level of accuracy and detail. The principle behind the BTE is the simple idea of charge conservation. Each particle, in the present case either electron or phonon, is assumed to occupy a spatial and

*Nanophononics: Thermal Generation, Transport, and Conversion at the Nanoscale*
Edited by Zlatan Aksamija

ISBN 978-981-4774-41-3 (Hardcover), 978-1-315-10822-3 (eBook)
www.panstanford.com

a momentum coordinate, which is the classical aspect of the BTE. A distribution function then counts the number of particles occupying each set of coordinates in space and momentum. Then conservation of particles in both space and momentum is enforced by equating the total rate of change in time to the change of the distribution function due to various scattering mechanisms.

## 2.2 The Boltzmann Transport Equation for Electrons

In the semiclassical formulation, particle position and momentum are both independent and functions only of time, so we can expand the gradient and apply the chain rule (Eq. 2.1).

$$\frac{df(x,k,t)}{dt}=\frac{\partial f}{\partial t}+\frac{\partial f}{\partial x}\frac{dx}{dt}+\frac{\partial f}{\partial k}\frac{dk}{dt}=\left(\frac{df}{dt}\right)\Big|_{\text{scat}}. \tag{2.1}$$

Furthermore, we can identify the derivative of position as the electron velocity $dx/dt = v(k)$. Velocity is computed from the gradient of the electronic band structure (Eq. 2.2).

$$v(k)=\frac{1}{\hbar}\frac{dE(k)}{dk}. \tag{2.2}$$

The derivative of momentum can be identified as the effect of the field

$$\frac{dk}{dt}=\frac{eF(x)}{\hbar},$$

where $e$ is the electron charge and $F(x)$ is the electric field strength at a given position. Finally, we obtain the final form of the BTE (Eq. 2.3), which is exactly the BTE that can be derived from more detailed considerations of electron conservation [1].

$$\frac{\partial f}{\partial t}+v(k)\frac{\partial f}{\partial x}+\frac{eF(x)}{\hbar}\frac{\partial f}{\partial k}=\delta f\,|_{\text{scat}}. \tag{2.3}$$

This form shows us that we have a first-order system of partial differential equations to solve in dimensions of time, space, and momentum. The advantage of solving for a distribution function in this formulation is that we can now obtain the total charge along the tube quite simply by integrating the distribution function over the momentum variable and multiplying by the electron charge (Eq. 2.4).

$$\rho(t,x) = e\int f(x,k,t)dk. \tag{2.4}$$

The charge density can now be used to determine the strength of the electric field along the tube. Similarly, the current can be determined by integrating the velocities of all the momentum states weighted by their distribution (Eq. 2.5).

$$I(x,t) = e\int v(k)f(x,k,t)dk. \tag{2.5}$$

Equation 2.5 allows us to obtain a value for the current once the steady-state distribution function $f(x, k, t)$ is known. The coupling of external applied potentials and the electronic transport in the nanotube is achieved through Poisson's equation (Eq. 2.6).

$$\nabla^2 V(x) = \frac{d^2V(x)}{dx^2} = \frac{\rho(x)}{\epsilon}. \tag{2.6}$$

## 2.3 Electron Scattering Rates

As shown in Eq. 2.3, the forcing term of the equation is the change in the distribution function due to collisions. Electrons can interact with quantized physical vibrations, called phonons, as well as impurities and other electrons in the nanotube lattice. In the absence of scattering, the distribution would advect in space and momentum under the influence of electron velocity and applied field. Scattering serves to limit this process and bring the distribution function to a steady-state value after some initial transient time. The scattering integral can be derived by considering the probabilities of transitions from one momentum state $k$ to another state $k'$, represented by $S(k, k')$ and reverse $S(k', k)$. The probability of occupying those states is given by the distribution functions $f(k)$ and $f(k')$. Then the scattering-out term is $S(k, k')f(k)$ and the scattering-into of a state $k$ is $S(k', k)f(k')$. Taking the difference and integrating over all the states $k'$ gives the total change in the distribution function due to collisions (Eq. 2.7).

$$\frac{df(x,k,t)}{dt} = \int dk'\left[S(k,k')f(x,k,t) - S(k',k)f(x,k',t)\right]. \tag{2.7}$$

The scattering probabilities $S(k, k')$ can be derived from quantum-mechanical perturbation theory, thereby including the correct scattering statistics [1]. We can also extend this calculation to include Pauli's exclusion principle, which states that no more than one electron can occupy the same quantum-mechanical state. This translates directly into an additional factor that counts the probability of a final state not being available, represented by $(1 - f(k))$. Inserting this factor complicates slightly the expression for the scattering integral (Eq. 2.8).

$$\delta f|_{\text{scat}} = \int dk'[S(k,k')f(k)(1-f(k')) - S(k',k)(1-f(k))f(k')]. \quad (2.8)$$

A slightly simpler form can be derived from Eq. 2.7 by assuming that the distribution function is only slightly perturbed from equilibrium by a small function, which we label $\delta f(x, k, t)$ (Eq. 2.9).

$$\frac{d}{dt}(f_{\text{eq}} + \delta f) = \int dk'[S(k,k')(f_{\text{eq}} + \delta f) - S(k',k)(f_{\text{eq}}' + \delta f')]. \quad (2.9)$$

Since the equilibrium distribution, given by the familiar Fermi–Dirac statistics (Eq. 2.10), is in a steady state, the overall scattering integral in equilibrium is zero (Eq. 2.11) [2]. This principle is referred to as detailed balance, since each process is exactly balanced out by an opposite but equal counterpart, causing no net change to the distribution function.

$$f_{\text{eq}}(x,k,t) = \left[\exp\left(\frac{E(k) - E_{\text{F}}(x)}{k_{\text{B}}T} + 1\right)\right]^{-1}. \quad (2.10)$$

$$\frac{df_{\text{eq}}(x,k,t)}{dt} = \int dk'[S(k,k')f_{\text{eq}}(x,k,t) - S(k',k)f_{\text{eq}}(x,k',t)] = 0. \quad (2.11)$$

This leaves only an expression involving $\delta f(x, k, t)$ (Eq. 2.12).

$$\frac{d\delta f(x,k,t)}{dt} = \int dk'[S(k,k')\delta f(x,k,t) - S(k',k)\delta f(x,k',t)]. \quad (2.12)$$

The perturbed portion, $\delta f$, is an odd function of momentum $\delta f(x, k, t) = -\delta f(x, -k, t)$, while the scattering probabilities are even functions $S(k, k') = S(-k, k')$, so the second term in the integral cancels, leaving us with a simplified form (Eq. 2.13).

$$\frac{d\delta f(x,k,t)}{dt} = \delta f(x,k,t)\int S(k,k')dk'. \tag{2.13}$$

The integral over all the scattering probabilities produces the total scattering rate $\Gamma(k)$, which is the inverse of the momentum relaxation time $\tau_{\mathrm{m}}(k)$ (Eq. 2.14).

$$\Gamma(k) = \frac{1}{\tau_{\mathrm{m}}(k)} = \int S(k,k')dk'. \tag{2.14}$$

To first order, this probability would be calculated from Fermi's golden rule [1, 3, 4], where $H$ is the matrix element and $\omega(q)$ is the phonon frequency. The minus and plus signs refer to emission and absorption, respectively. Equation 2.15 is simplified somewhat by using the rigid pseudo-ion approximation [5] for the matrix element.

$$\begin{aligned} S(k,k') &= \frac{2\pi}{\hbar}|H(k,k')|^2\,\delta(E(k)-E(k')\pm\hbar\omega(q)) \\ &= \frac{2\pi}{\hbar}\frac{|D(q)|^2\,I^2(q)}{\rho\omega(q)}\left(N(x,q,t)+\frac{1}{2}\pm\frac{1}{2}\right)\delta(E(k)-E(k')\pm\hbar\omega(q)) \end{aligned} \tag{2.15}$$

Implementing this formula requires numerical integration over the whole momentum space and use of the complete phonon and electron dispersion relationships. This would then be repeated on some fine grid, and for each point we would have a numerical integration to perform [6, 7]. This would give us a probability to scatter from each position in momentum space.

$$\frac{df(x,k,t)}{dt} = \frac{\delta f(x,k,t)}{\tau_{\mathrm{m}}(k)} = \frac{f(x,k,t)-f_{\mathrm{eq}}(x,k,t)}{\tau_{\mathrm{m}}(k)}. \tag{2.16}$$

We can see from Eq. 2.16 that any disturbance to the equilibrium distribution relaxes back to equilibrium with a time constant equal to the relaxation time. The time-dependent solution of the BTE then represents a balance between drift and diffusion, acting to displace the distribution function from equilibrium, and scattering acting to return it to equilibrium through the relaxation time. The relaxation time can be precomputed by appropriate methods and stored for use in the simulation.

## 2.4 The Phonon Boltzmann Transport Equation

The principle of conservation of particles applies equally to phonons. Therefore, the same equation governs their transport, with some modifications. The key difference between electrons and phonons is that phonons have no charge and, therefore, do not interact directly with each other or with any applied potentials. Their only interaction is through the atomic potential, leading to spontaneous decay if the potential is not perfectly quadratic, a property often termed "anharmonic" [8]. The phonon BTE will, therefore, have one less term on the left-hand side of Eq. 2.17 since the only change of the distribution function is due to the motion of phonons in real space.

$$\frac{\partial N(x,q,t)}{\partial t}+v(q)\frac{\partial N(x,q,t)}{\partial x}=\delta N(x,q,t)|_{\text{el-ph}}-\delta N(x,q,t)|_{\text{anharm}}. \tag{2.17}$$

The right-hand side of the phonon BTE is considerably more involved than its electron counterpart. The phonon distribution function, usually notated by $N\ (x,\ q,\ t)$, increases every time an electron-phonon scattering event takes place and decreases for every anharmonic phonon decay. Therefore, we will have two terms, one owing to the interactions with electrons and the other due to anharmonicity. The first term is derived by rearranging the corresponding term for electrons. We notice that for every transition an electron makes from an initial state $k$ to a final state $k'$, a phonon with a momentum $q = \pm(k - k')$ is either absorbed or emitted, depending on the energies of the two electron states. Every emission increases the phonon occupancy and adds one to $N\ (x,\ q,\ t)$, while every phonon absorption does the opposite, and decreases $N\ (x,\ q,\ t)$ by unity [9]. The electron states are weighted by their probability of occupancy, given by the electron distribution function $f\ (x,\ k,\ t)$. The total rate of change of the phonon distribution function is obtained by summing up all the electron states, as in Eq. 2.18.

$$\begin{aligned}\frac{dN(q)}{dt} = \frac{2\pi}{\Omega}\int dk\frac{|D(q)|^2\ I^2(q)}{\rho\hbar\omega(q)}\\ \{(N(x,q,t)+1)f(x,k\pm q,t)\delta(E(k\pm q)-E(k)\pm\hbar\omega(q))\\ -N(x,q,t)f(x,k,t)\delta(E(k)-E(k\pm q)\pm\hbar\omega(q))\}\end{aligned} \tag{2.18}$$

The second term in the scattering portion of the phonon BTE (Eq. 2.17) is due to coupling with other phonon modes through the anharmonic part of the atomic potential. This gives rise to a decay process for phonons that is especially important in optical branches, which, due to their higher energies, have ample opportunity to decay into pairs of acoustic phonons. Because such processes involve three phonons, they are also known as three-phonon decay or scattering. There are also higher-order processes involving four or more phonons, but these are less directly applicable to our study. Also, owing to the higher order of a four- and five-phonon process, the strength of the interaction is smaller by an order of magnitude or more because the atomic potential is smooth and does not give rise to significant quadric or quintic terms.

## 2.5 Phonon Scattering and Anharmonic Decay

The anharmonic phonon decay can be thought of as very similar to the electron-phonon interaction, in the sense that we view the scattering event as one phonon entering with an initial state $q$ and two phonons exiting with final states $q'$ and $q \pm q'$. Energy and momentum are conserved, governed by the expressions $q = q' \pm q''$ and $\omega(q) = \omega(q') \pm \omega(q \pm q')$, similar to the electron case. This produces an expression for the rate of change of the phonon distribution function (Eq. 2.19) that depends on the distributions at the final states in a nonlinear fashion, again much like the electron-phonon interaction.

$$\begin{aligned}\frac{dN(x,q,t)}{dt} = \frac{2\pi}{\Omega}\int dq' \frac{c^2(q,q',q\pm q')}{\rho\hbar\omega(q)\omega(q')\omega(q\pm q')}\delta(\omega(q)-\omega(q')\pm\omega(q\pm q'))\\ \{N(x,q,t)(N(x,q',t)+1)(N(x,q\pm q',t)+1)\\ -(N(x,q,t)+1)N(x,q',t)N(x,q\pm q',t)\}\end{aligned} \tag{2.19}$$

The interaction potential is derived from the cubic portion of the atomic potential but can be simplified considerably by assuming it depends only on an average parameter $\gamma$ called the *Grüneisen* constant, which is determined experimentally from thermal expansion and compressibility [8]. This allows a simple relationship to be derived for the anharmonic coupling potential (Eq. 2.20).

$$c(q,q',q\pm q') = -\frac{i}{\sqrt{N}}\frac{2M}{\sqrt{3}v(q)}\gamma\,\omega(q)\omega(q')\omega(q\pm q'). \tag{2.20}$$

As for the electron case, a relaxation time expression can be derived for phonons from the BTE (Eq. 2.18) by considering only a perturbation $\delta N$ to the equilibrium distribution function.

$$\begin{aligned}\frac{d(\delta N)}{dt}=\frac{2\pi}{\Omega}\int dq'\frac{4M^2\gamma}{3Nv(q)}\omega(q)\omega(q')\omega(q\pm q')\\ \{(N_{\text{eq}}(x,q,t)+\delta N)(N_{\text{eq}}(x,q',t)+1)(N_{\text{eq}}(x,q\pm q',t)+1)\\ -(N_{\text{eq}}(x,q,t)+\delta N+1)N_{\text{eq}}(x,q',t)N_{\text{eq}}(x,q\pm q',t)\}\\ \delta(\omega(q)-\omega(q')\pm\omega(q\pm q'))\end{aligned} \tag{2.21}$$

The expression in Eq. 2.21 is simplified by considering the equilibrium condition (Eq. 2.22), again based on detailed balance.

$$\begin{aligned}\int dq'\{N_{\text{eq}}(x,q,t)(N_{\text{eq}}(x,q',t)+1)(N_{\text{eq}}(x,q\pm q',t)+1)\\ -(N_{\text{eq}}(x,q,t)+1)N_{\text{eq}}(x,q',t)N_{\text{eq}}(x,q\pm q',t)\}\\ \delta(\omega(q)-\omega(q')\pm\omega(q\pm q'))=0\end{aligned} \tag{2.22}$$

Finally we arrive at a relaxation time expression for phonons (Eq. 2.23) that is valid as long as the final modes are near equilibrium.

$$\begin{aligned}\frac{dN(x,q,t)}{dt}&=\frac{N(x,q,t)-N_{\text{eq}}(x,q,t)}{\tau(q)}\\ &=\frac{2\pi}{\Omega}\int dq'\frac{4M^2\gamma}{3Nv(q)}\omega(q)\omega(q')\omega(q\pm q')\\ &\{\delta N(x,q,t)(N(x,q',t)+1)(N(x,q\pm q',t)+1)\\ &-\delta N(x,q,t)N(x,q',t)N(x,q\pm q',t)\}\\ &\delta(\omega(q)-\omega(q')\pm\omega(q\pm q'))\end{aligned} \tag{2.23}$$

From Eq. 2.23 we can factor out the perturbed portion $\delta N$ and divide by it in order to obtain an expression for the three-phonon scattering rate, or the inverse of the relaxation time (Eq. 2.24).

$$\begin{aligned}\frac{1}{\tau(q)}&=\frac{d\delta N(x,q,t)}{dt}\frac{1}{\delta N(x,q,t)}\\ &=\frac{2\pi}{\Omega}\int dq'\frac{4M^2\gamma\omega(q)\omega(q')\omega(q\pm q')}{3Nv(q)}\{N(x,q',t)+N(x,q\pm q',t)+1\}\\ &\delta(\omega(q)-\omega(q')\pm\omega(q\pm q'))\end{aligned} \tag{2.24}$$

This integral can be evaluated numerically on a grid in momentum space and the relaxation time stored for use in simulation.

## References


1. Hess, K. (2000). *Advanced Theory of Semiconductor Devices*, IEEE Press, New York, NY.
2. Kittel, C. (2005). *Introduction to Solid State Physics*, John Wiley and Sons, New York, NY.
3. Nguyen, P. H., Hofmann, K. R., and Paasch, G. (2003). Comparative full-band Monte Carlo study of Si and Ge with screened pseudopotential-based phonon scattering rates, J. Appl. Phys., **94** (1), pp. 375–386.
4. Jacoboni, C., and Reggiani, L. (1983). The Monte Carlo method for the solution of charge transport in semiconductors with applications to covalent materials, *Rev. Mod. Phys.*, **55**(3), pp. 645–705.
5. Nguyen, P. H., Hofmann, K. R., and Paasch, G. (2002). Full-band Monte Carlo model with screened pseudopotential based phonon scattering rates for a lattice with basis, *J. Appl. Phys.*, **92**(9), pp. 5359–5370.
6. Kunikiyo, T., Takenaka, M., Kamakura, Y., Yamaji, M., Mizuno, H., Morifuji, M., Taniguchi, K., and Hamaguchi, C. (1994). A Monte Carlo simulation of anisotropic electron transport in silicon including full band structure and anisotropic impact-ionization model, *J. Appl. Phys.*, **75**(1), pp. 297–312.
7. Yoder, P. D., Higman, J. M., Bude, J., and Hess, K. (1992). Monte Carlo simulation of hot electron transport in Si using a unified pseudopotential description of the crystal, *Semicond. Sci. Technol.*, **7**, pp. B357–B359.
8. Seitz, F., and Turnbull, D. (eds.) (1970). *Solid State Physics*, Vol. 7, Academic Press, New York, NY.
9. Conwell, E. M. (1967). *High Field Transport in Semiconductors*, Academic Press, New York, NY.

# Chapter 3

# Phonon Emission and Absorption Spectra in Silicon

**Zlatan Aksamija**
*University of Massachusetts Amherst, Amherst, MA 01003, USA*
zlatana@engin.umass.edu

## 3.1 Introduction

The thermal budget is quickly emerging as the prominent limitation on future trends in scaling of semiconductor devices. A detailed understanding of electron-phonon coupling, as well as characteristics of the phonon heat generated by the electron current, such as the phonon mode and spectrum, is crucial to our understanding of the microscale heating issues in semiconductor devices [1]. High electric fields present in ultrascaled devices pose additional challenges to the understanding of electrothermal properties. Identifying trends in phonon emission spectra at high fields is crucial to enabling more thermally efficient devices. Previous work in phonon emission in silicon relied on the Monte Carlo simulation of electron transport. Such simulations capture the electron-phonon interaction for the

*Nanophononics: Thermal Generation, Transport, and Conversion at the Nanoscale*
Edited by Zlatan Aksamija

ISBN 978-981-4774-41-3 (Hardcover), 978-1-315-10822-3 (eBook)
www.panstanford.com

purpose of accurately describing the electron current and usually simplify the treatment of phonons by assuming a simple analytic dispersion relationship between phonon momentum and energy $\hbar\omega(\mathbf{q})$ [2, 3]. This is usually sufficient to describe electron transport with great accuracy. Such Monte Carlo simulations then keep track of the interactions between electrons and phonons, from which the number of phonons generated during the course of the simulation can be extracted and the relationship between the net phonon generation rate and the phonon energy can be computed. This was done by Pop et al. [4] for silicon and strained silicon on the basis of analytical isotropic phonon dispersion and analytical multivalley nonparabolic electron band structure. Although the main trends could be identified, lack of a full electron band structure, and even more importantly the use of analytical isotropic phonon dispersion, did not capture the full detail of the phonon emission spectra. In addition, the Monte Carlo method, due to its stochastic nature, requires a large number of particles and iterations for convergence. To avoid these limitations, we use a deterministic method given by Gilat and Raubenheimer and extend it to calculate the rate of change of the phonon distribution function due to electron-phonon interactions.

## 3.2 The Adiabatic Bond Charge Model for Phonons

The relationship between phonon energy and momentum is usually presented by the phonon dispersion relationship. This relationship bears some similarity to the usual electron dispersion, better known as the band diagram. A phonon, much like an electron, can take on any momentum in the Brillouin zone. Unlike the electron, phonon energies have a much smaller range, going from 0 meV to about 65 meV. This is a continuum of energies. For any energy in this range, there is a momentum corresponding to it rather than the usual energy gap we note in the electron band diagram. There are six branches on this dispersion. The primary classification of phonons is by energy or, correspondingly, frequency. They are bosons because they obey Bose–Einstein statistics (Eq. 3.1) for the average occupation number at a given mode [4].

$$N(\mathbf{q}) = \left[ \exp\left( \frac{\hbar\omega(\mathbf{q})}{k_B T_{ph}} \right) - 1 \right]^{-1} \tag{3.1}$$

Their frequency and energy also follow the simple relationship $E = hf$ or equivalently $E = \hbar\omega$. Therefore, low energies near zero correspond to low frequencies, even frequencies in the range of human hearing (20 – 20 000 Hz), so these phonons are called acoustic. Others can have frequencies in the range of optical light, enabling them to interact with photons, and therefore, they are called optical. This classification is depicted in Fig. 3.1. Another distinction between acoustic and optical modes is in the mode of vibration. Since most semiconductors of interest have two atoms per basis, each with 3 degrees of freedom, each atom in the basis contributes three phonon branches. Acoustic phonons give rise to the two atoms vibrating in phase with each other, while optical phonons represent standing waves where the two atoms vibrate against one another [5]. At the edge of the Brillouin zone, the two meet in the case of silicon and germanium because the two atoms in the basis have equal mass, while in III–V and II–IV compounds, there exists a gap between acoustic and optical branches owing to the difference in mass of the two different atoms.

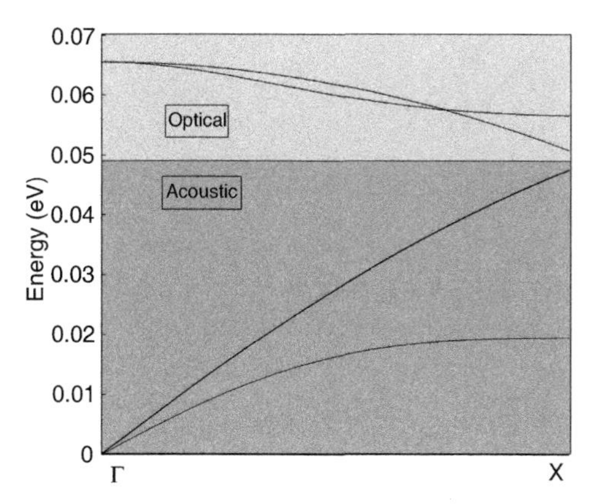


**Figure 3.1** Classification of phonon modes into optical and acoustic according to their frequencies.

The dispersion used in this project was calculated using an adiabatic bond charge (ABC) model [6–8]. This model was chosen because it gives results within 2% of measured values, without the computational burden of ab initio methods [9], and can be taken as reliable for our purposes [10, 11]. The model was suggested nearly three decades ago as an alternative to the similar valence shell models. The noted advantage of the ABC model was that it had only four parameters that were variable, so it was less prone to error and more easily calibrated to give accurate results [7]. The model itself consists of representing the lattice vibrations as 3D forces between the four types of interactions:

- Nearest-neighbor ion
- Bond-charge
- Coulomb
- Bond–bond

These four interactions are presented schematically in Fig. 3.2.

The results of the ABC model are shown in Fig. 3.3. They are further distinguished by the direction of vibration of the lattice relative to the direction of propagation, leading to longitudinal and transverse modes. Furthermore, there is degeneracy in the transverse branches, so we have in total six phonon types, listed in Table 3.1.

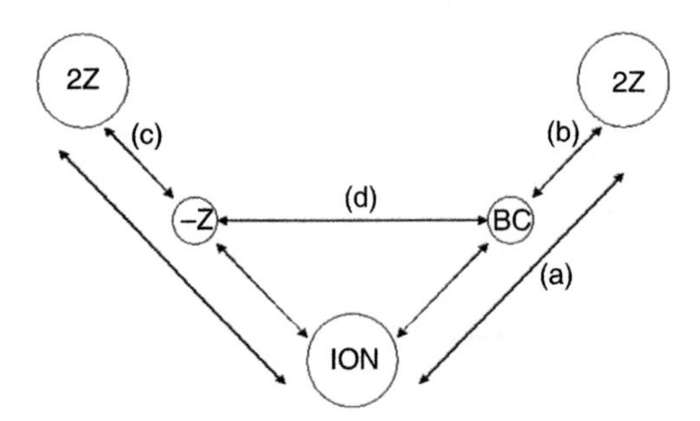


**Figure 3.2** Schematic presentation of the adiabatic bond charge model depicting the four types of interactions included in the model.

**Table 3.1** Six phonon types due to degeneracy

| Branch | Frequency | Direction | Degeneracy | Short name |
|---|---|---|---|---|
| 1 | Optical | Transverse | branch 1 | TO1 |
| 2 | | | branch 2 | TO2 |
| 3 | | Longitudinal | none | LO |
| 4 | Acoustic | Longitudinal | none | LA |
| 5 | | Transverse | branch 1 | TA1 |
| 6 | | | branch 2 | TA2 |

## 3.3 Numerical Computation of Phonon Spectra

The rate of change of the phonon distribution can be computed by integrating the scattering rate over all electron momenta, as in Eq. 3.3 [12]. In this work, we strive to combine established numerical algorithms for Brillouin zone integration [13] with the deformation potentials given in the literature [3] to compute detailed electron-phonon coupling and phonon absorption and emission rates. The electronic band structure is obtained from the empirical pseudopotential method [14]. To capture the band structure more accurately, our implementation was extended to include nonlocal pseudopotentials [15], while Weber's adiabatic bond-charge model is used for phonon dispersion [7]. This model allows a very accurate representation of all phonon branches, shown in Fig. 3.3, even along directions other than the main symmetry directions [8].

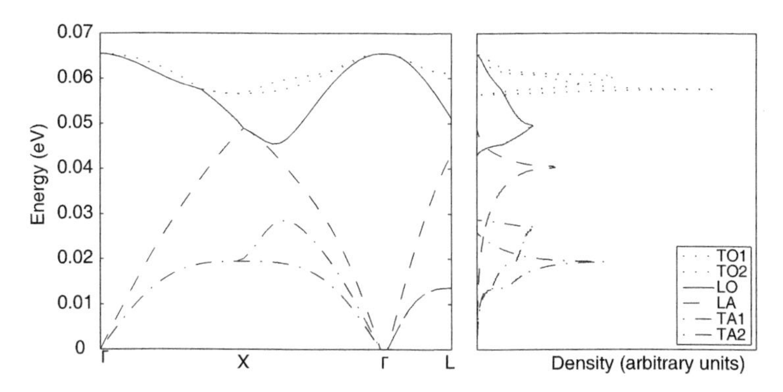


**Figure 3.3** Phonon dispersion relationship (left) and phonon density of states (DOS) for each of the six phonon modes and total (right). The adiabatic bond charge model is used to compute the phonon dispersion on a regular grid in momentum space. The phonon generation spectrum is found to be similar to the DOS, which is dominated by optical and longitudinal acoustic (LA) modes. The DOS also shows that the optical modes are found above 48 meV, transverse acoustic (TA) modes are below 25 meV, and LA modes are in between.

Phonon transport is described by the Boltzmann transport equation (Eq. 3.2), which describes the evolution in time of the phonon distribution function. This distribution function, denoted by $N(\mathbf{r}, \mathbf{q}, t)$, represents the average number of phonons that occupy a state $\mathbf{q}$ in position $\mathbf{r}$ and at time $t$. The rate of change of the phonon distribution has several contributions. One of the dominant contributions is due to the anharmonic phonon processes, which arise due to third- and higher-order components of the bond forces. This is a phonon-phonon interaction, which allows the phonon population to return to equilibrium when perturbed by a temperature gradient or a generation source. A second contribution that is strongly present in semiconductor devices is due to the generation of phonons by the electron-phonon interaction. This interaction occurs when strong electric fields drive the electron population out of equilibrium and impart energy to the electrons. The electron-phonon interaction is the primary means for electrons to release the energy accumulated from the accelerating effect of the applied electric field and to return to equilibrium. This is especially important in nanoscale transistor devices, where electrons are accelerated by the strong fields in the channel region of the transistor, and then the electrons have to release energy in the form of phonons in order to return to quasi-equilibrium in the drain region before they leave the transistor through the contacts. Generation, as well as absorption, of phonons couples together the electron and phonon populations and allows energy exchange between them, leading to a net flow of energy from the electron gas to the phonon bath. The phonon population under the heating effect of electrons can be described by the phonon Boltzmann transport equation with the added term due to the electron-phonon interaction [16]. The phonon distribution function, notated by $N(\mathbf{r}, \mathbf{q}, t)$, increases every time an electron-phonon scattering event takes place and decreases for every anharmonic phonon decay. Therefore, we will have two terms on the right-hand side, one owing to the interactions of phonons with electrons and the other due to interactions of phonons with other phonons through anharmonic decay, boundary scattering, isotope scattering, and other interaction mechanisms, which are beyond the scope of this work.

$$\frac{\partial N(\mathbf{r},\mathbf{q},t)}{\partial t} + v(q)\nabla_{\mathbf{r}} N(\mathbf{r},\mathbf{q},t) = \left(\frac{dN(\mathbf{r},\mathbf{q},t)}{dt}\right)_{\text{el-ph}} - \left(\frac{dN(\mathbf{r},\mathbf{q},t)}{dt}\right)_{\text{ph-ph}} \tag{3.2}$$

In considering this challenging transport equation, we will restrict our attention to homogenous materials in the steady state. In the steady state, all variation with time has vanished, so the first

term containing a partial derivative with respect to time can be removed. The homogenous restriction means we are considering the case of bulk material, so all gradients with respect to the spatial coordinate are zero. This removes from consideration the second term of the phonon Boltzmann transport equation, containing spatial gradients of the phonon distribution function. Finally, the term due to anharmonic decay is usually treated in the relaxation time approximation, and the relaxation rates have been presented by several researchers [17, 18] on the basis of analytical models of phonon dispersion, as well as detailed numerical computations of decay rates of optical phonons on the basis of an empirical model of phonon dispersion [19] and density functional theory (DFT) [20]. On the other hand, the generation term due to electron-phonon interaction has not been given the same amount of attention, as it does not play a significant role in purely thermal transport and contributes little when there are no electric fields present, because the electron population is in equilibrium with phonons. More recently, interest in coupled electrothermal simulation has grown [21, 22] due to inherent thermal limitations to electron transport. It is becoming necessary to model accurately the response of phonons to heating due to the release of thermal energy through electron-phonon interaction. This is significant when electrons are subjected to very high electric fields [2, 4], such as those that are present in the channel region of semiconductor devices.

The rate of change of the phonon distribution function due to electron-phonon scattering can be computed by combining the contributions from the electron-phonon scattering over all electron momenta. We notice that for every transition an electron makes from an initial state $k$ to a final state $k'$, a phonon with a momentum $\mathbf{q} = \pm(k - k')$ is either absorbed or emitted, depending on the energies of the two electron states. Every emission increases the phonon occupancy and adds one to $N(\mathbf{q})$, while every phonon absorption does the opposite and decreases $N(\mathbf{q})$ by unity [23]. The electron states are weighted by their probability of occupancy, given by the electron distribution function $f_e(k)$. The total rate of change of the phonon distribution function is obtained by summing over all the electron states [24], as in Eq. 3.3.

$$\left(\frac{dN(\mathbf{q})}{dt}\right)_{\text{el-ph}} = \sum_{\mathbf{k}} \frac{|D(\mathbf{q})|^2 I^2(q)}{\rho\omega(\mathbf{q})}\left(N(\mathbf{q}) + \frac{1}{2} \pm \frac{1}{2}\right) \quad (3.3)$$
$$f_{\text{e}}(\mathbf{k})\delta(E(\mathbf{k}) - E(\mathbf{k} \pm \mathbf{q}) \pm \hbar\omega(\mathbf{q}))$$

In this work, we are interested in bulk silicon crystal, which is assumed to be infinite in extent. This implies that momentum states are also infinite in number, so they form a continuum. Then the summation over all momentum states becomes an integral over the entire first Brillouin zone for the purpose of calculation [5]. Such an integral over the entire first Brillouin zone can be computed numerically by dividing the Brillouin zone into a regular mesh of small cubes or tetrahedra and computing the area of intersection of the energy isosurface with each particular cube or tetrahedron. There are several algorithms that are well suited for the purpose of such a calculation. We choose the algorithm given by Gilat and Raubenheimer, which uses a division of the momentum space into regular small cubes, expanding all the terms in the integrand to first order for the purpose of obtaining a very accurate result and taking into account the strong gradients in electron energy and distribution functions. Due to the inclusion of terms up to the linear in **q**, that is, the first derivative of energy $E(\mathbf{q})$, in the calculation, this approach was also termed "linear analytic" [13]. This algorithm was first proposed for the purpose of calculating spectral properties, such as the density of states (DOS). More recently, the Gilat–Raubenheimer (GR) approach was applied to a very accurate calculation of the electron-phonon scattering rates in silicon and other semiconductors by Fischetti and Laux [3]. Unlike the work of Fischetti and Laux, who used an analytical expression for the dispersion relationship of phonons, we use the full dispersion computed from Weber's ABC model in order to accurately obtain the dependance on momentum of the phonon energy $\hbar\omega(\mathbf{q})$ in the integrand of Eq. 3.4.

$$\left(\frac{dN(\mathbf{q})}{dt}\right)_{\text{el-ph}} = \frac{2\pi}{\Omega}\int d\mathbf{k} \frac{|D(\mathbf{q})|^2 I^2(q)}{\rho\omega(\mathbf{q})}\left(N(\mathbf{q}) + \frac{1}{2} \pm \frac{1}{2}\right) \quad (3.4)$$
$$f_{\text{e}}(\mathbf{k})\delta(E(\mathbf{k}) - E(\mathbf{k} \pm \mathbf{q}) \pm \hbar\omega(\mathbf{q}))$$

Energy conservation is expressed by the $\delta$-function, which defines a complex energy-conserving surface in the first Brillouin zone. Using properties of the $\delta$-function, the integral over **k** is converted into a surface integral over the entire energy-conserving surface (Eq. 3.5). This surface is composed of all the final momentum states $k'$ that are able to conserve both momentum and energy with a given initial state $k$ according to $\Delta E = 0$, where $\Delta E = E(\mathbf{k}) - E(\mathbf{k} \pm \mathbf{q}) \pm \hbar\omega(\mathbf{q})$. Then the contribution to the integral from each small cube in momentum space is given by the area of the intersection between this energy-conserving surface and that particular cube, weighted in our case by the contribution from the matrix element.

$$\left(\frac{dN(\mathbf{q})}{dt}\right)_{\text{el-ph}} = \frac{2\pi}{\Omega}\oint_{\Delta E=0} \frac{d\mathbf{S}}{|\nabla_{\mathbf{k}}E(\mathbf{k}) - \nabla_{\mathbf{k}}E(\mathbf{k}\pm\mathbf{q})|} \frac{|D(\mathbf{q})|^2\, I^2(q)}{\rho\omega(\mathbf{q})}\left(N(\mathbf{q}) + \frac{1}{2} \pm \frac{1}{2}\right) f_{\mathrm{e}}(\mathbf{k})$$

In the GR algorithm, the portion of the energy-conserving surface in each cube is assumed to be a plane, so the shape of the intersection of a plane and a cube then becomes a triangle, a quadrangle, or a pentangle, depending on the angle of the gradients of the electron and phonon energy dispersion relationships in each particular cube. The contribution $S_i$ arising from each small cube centered at $\mathbf{k}_i$ and indexed by $i$ in the entire Brillouin zone can then be computed on the basis of geometric considerations and summed together to obtain the total rate of change of the phonon distribution function $N(\mathbf{q})$, as given in Eq. 3.5. Further details of the GR *linear analytic* algorithm and the expressions for the areas of intersections of equal energy surfaces with cubes in momentum space are given in Ref. [25]. This algorithm is very accurate and well-suited to using tabulated data for band structures and dispersions [13].

$$\left(\frac{dN(\mathbf{q})}{dt}\right)_{\text{el-ph}} = \frac{|D(\mathbf{q})|^2\, I^2(q)}{\rho\omega(\mathbf{q})}\left(N(\mathbf{q}) + \frac{1}{2} \pm \frac{1}{2}\right)\sum_i S_i(\mathbf{k}_i) f_{\mathrm{e}}(\mathbf{k}_i)$$

The deformation potential for acoustic phonon interaction, due to the crystal symmetry, reduces to two independent parameters, the dilatation potential $\Xi_d$ and the uniaxial shear potential $\Xi_u$ [23, 26]. The relationship between the deformation potential and these two parameters is described by the Herring–Vogt relation [27], which expresses the acoustic deformation potential in terms of the dilatation and shear potentials, along with an angular dependence. The angular dependence is on $\Theta$, the angle between the phonon momentum $\mathbf{q}$ and the longitudinal axis of the conduction band valley [27]. The acoustic deformation potential also depends on the phonon polarization, longitudinal or transverse, and is expressed in Eq. 3.5:

$$\begin{aligned} D_{\mathrm{LA}}(\mathbf{q}) &= (\Xi_d + \Xi_u \cos^2\Theta)\mathbf{q} \\ D_{\mathrm{TA}}(\mathbf{q}) &= (\Xi_u \sin\Theta\cos\Theta)\mathbf{q} \end{aligned} \tag{3.5}$$

For optical phonons, the interaction is of the zeroth order and the deformation potential, expressed in Eq. 3.6, is taken to be a constant with values taken from Ref. [3]. The overlap integral between the cell portions of the Bloch states [27] is computed in the Nordheim, or spherical cell, approximation [28]. This provides an analytical expression for the integral (Eq. 3.7), which depends only on the magnitude of momentum of the phonon being exchanged, $\mathbf{q} = /k - k'/$. The $R_s = a_{\mathrm{Si}}\,[3/(16\pi)]^{1/3}$ is the radius of the Wigner–Seitz cell.

$$D_{\mathrm{LO,TO}}(\mathbf{q}) = DK_{\mathrm{op}} \tag{3.6}$$

$$I(\mathbf{q}) = \frac{3}{(\mathbf{q}R_s)^3}\left[\sin(\mathbf{q}R_s) - \mathbf{q}R_s\cos(\mathbf{q}R_s)\right] \tag{3.7}$$

The electron and phonon populations are assumed to be initially in equilibrium, with a Fermi distribution (Eq. 2.10) for electrons and a Bose–Einstein distribution (Eq. 3.1) for phonons [29]. The electric field is introduced through a full-band Monte Carlo simulation [30, 31]. The simulation converges to a nonequilibrium electron distribution, which includes the effect of the applied electric field. This provides an accurate distribution for electrons [32]. The effect of the applied electric field can then be understood through the concept of equivalent electron temperature $T_e$ [5]. Increasing the applied field imparts more energy on the electron population and pushes the electron temperature further up from the lattice phonon temperature $T_{ph}$. Fitting the semilogarithmic plot of the Monte Carlo results for the electron distributions at various applied electric fields, as shown in Fig. 3.4, allows us to extract the electron temperature from each distribution and relate it to the applied electric field. To account for the effect of the electric field on the electron, and ultimately the phonon population, the heated displaced Maxwellian distribution, shown in Eq. 3.8, is often used. This is well justified when the carrier density is high, because carrier-carrier interactions very effectively thermalize the distribution [33]. We also use a heated and displaced quasi-equilibrium distribution for electrons based on the Fermi–

Dirac distribution, with the appropriate electron temperature $T_e$ and displacement, or drift vector, $\mathbf{k}_d$ determined from fit-to-bulk full-band Monte Carlo results at each given electric field. This distribution, given in Eq. 3.9, is very similar to the heated displaced Maxwellian but is also capable of accounting for Pauli's exclusion principle, which is important in order to account for quantum effects at near-degenerate doping levels and carrier concentrations above $10^{20}$ per cubic centimeter [34], such as those often found in the source and drain regions of submicron silicon metal-oxide-semiconductor field-effect transistor (MOSFET) devices.

$$f_{\mathrm{HDM}}(\mathbf{k}) = \exp\left(-\frac{E(|\mathbf{k}-\mathbf{k}_\mathrm{d}|)-E_\mathrm{F}}{k_\mathrm{B}T_\mathrm{e}}\right) \tag{3.8}$$

$$f_{\mathrm{HDFD}}(\mathbf{k}) = \left[\exp\left(\frac{E(|\mathbf{k}-\mathbf{k}_\mathrm{d})-E_\mathrm{F}}{k_\mathrm{B}T_\mathrm{e}}\right)+1\right]^{-1} \tag{3.9}$$

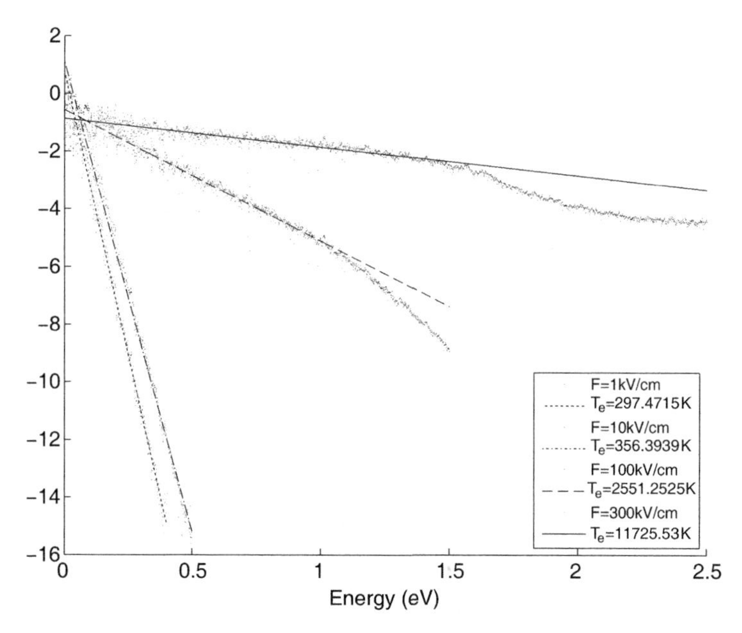


**Figure 3.4** Logarithmic plot of the electron distribution histogram. This figure demonstrates that the electron distributions remain linear over the range of energies of interest, even at high applied electric fields. This allows the definition of an equivalent electron temperature $T_e$, which reflects the heating of the electron population by the applied electric field.

The gradient of the electron energy can be large, and the accuracy of the Brillouin zone algorithm in Ref. [25] can be improved by expanding the electron distribution function up to first order and including the linear terms, as given in Ref. [35]. The gradient of the quasi-equilibrium electron distribution function (Eq. 2.10) can be determined analytically (Eq. 3.10), while the gradients of the electron energies can be computed numerically and tabulated along with the results for the electron energies. This improvement means we can incorporate the gradient into the summation in Eq. 3.5 in order to obtain a more accurate method, thereby improving the quality of the final results for phonon emission and absorption rates calculated from the electron distributions.

$$\begin{aligned}\nabla_{\mathbf{k}} f(\mathbf{k}) &= \frac{\partial f(E)}{\partial E} \nabla E(\mathbf{k}) \\ &= -\frac{1}{k_{\mathrm{B}} T_{\mathrm{e}}} f(\mathbf{k})(1 - f(\mathbf{k})) \nabla E(\mathbf{k})\end{aligned} \tag{3.10}$$

Because the integral in Eq. 3.5 is over all electron states $\mathbf{k}$, terms depending only on the phonon momentum $\mathbf{q}$ have been removed from the integral. In addition, due to its dependence on the phonon momentum $\mathbf{q}$ only, the phonon distribution function $N(\mathbf{q})$ can be factored out of the integral in Eq. 3.5 and divided by it to obtain a rate $\Gamma(\mathbf{q}) = 1/N(\mathbf{q})(dN(\mathbf{q})/dt)$. The dependence of this rate on the phonon energy $\omega$, rather than momentum $\mathbf{q}$, is obtained by averaging over all modes near a given energy $\omega$ using the same GR approach (Eq. 3.11). The phonon emission spectrum $\Gamma$ is then scaled to the phonon DOS, shown in Fig. 3.3·

$$\Gamma(\omega) = \frac{\int \Gamma(\mathbf{q}) \delta(\omega - \omega(\mathbf{q})) d\mathbf{q}}{\int \delta(\omega - \omega(\mathbf{q})) d\mathbf{q}} \tag{3.11}$$

This converts the distribution from momentum space into an energy spectrum $\Gamma(\omega)$ and enables us to examine the results. Each of the phonon types can be identified with a particular energy range. The DOS in Fig. 3.3 shows that the optical modes are found above 48 meV, transverse acoustic (TA) modes are below 25 meV, and longitudinal acoustic (LA) modes are in between. We can, therefore, examine trends in these three phonon types by looking at their respective energy ranges, low for TA, high for optical, and middle for LA phonons.

## 3.4 Results and Discussion

Due to the large differences in the directions and velocities of phonon propagation for different phonon modes, examining how much of each phonon polarization is generated is important for understanding heating in silicon at the microscale. Transverse optical modes have very flat dispersion curves, and consequently have the lowest velocities, as shown in Fig. 3.5. The longitudinal optical (LO) mode also tends to zero in the long-wavelength limit, but its velocity increases for short-wavelength phonons. The TA modes have larger velocities and approach the speed of sound in the long-wavelength limit. The highest velocities are present for the LA mode due to the fact that longitudinal waves propagate along the directions of the bonds in the crystal lattice. Phonon velocities can be averaged over all modes near a particular energy to obtain a velocity spectrum $v(\omega)$ that depends on phonon energy $\omega$ rather than the 3D momentum vector $\mathbf{q}$ and is, therefore, more suitable for plotting and comparisons. This is done by integrating phonon velocities over the

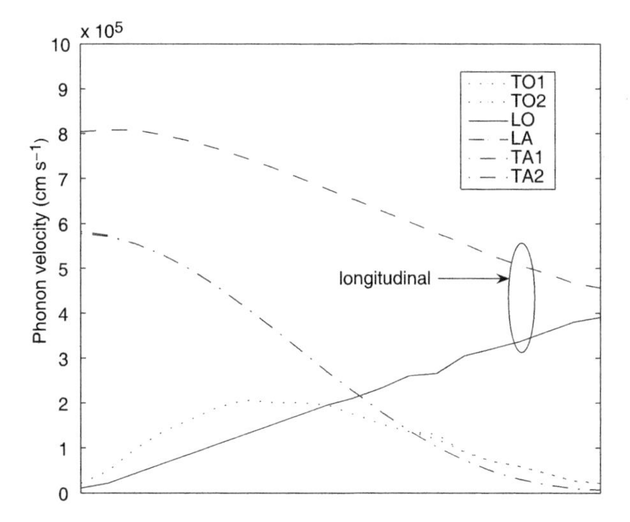


**Figure 3.5** Plot of phonon velocities. Group velocity is given by the gradient of the phonon dispersion relationship. Phonon velocities dictate how fast each mode can propagate heat through the silicon lattice. Longitudinal modes (dashed line), especially the longitudinal acoustic mode, have the highest velocity. Transverse acoustic modes are slower, while the optical modes are very slow due to their flat dispersion.

energy isosurface and then scaling by the DOS, as in Eq. 3.12. The velocity spectrum $v(\omega)$, shown in Fig. 3.6, demonstrates that phonon velocities for each mode are the highest at a low energy for phonons near the Brillouin zone center and lower for higher-energy zone-edge phonons.

$$v(\omega)=\frac{\int \nabla_{\mathbf{q}}\omega(\mathbf{q})\delta(\omega-\omega(\mathbf{q}))d\mathbf{q}}{\hbar\int \delta(\omega-\omega(\mathbf{q}))d\mathbf{q}} \tag{3.12}$$

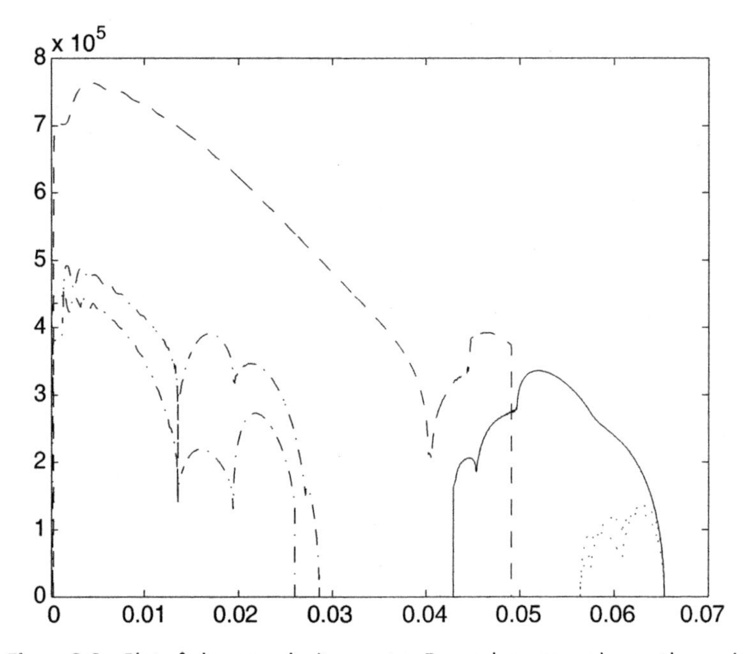


**Figure 3.6** Plot of phonon velocity spectra. For each energy value on the *x* axis, the velocity is computed by numerically averaging over the energy isosurface. This plot shows the average phonon velocity at each particular energy rather than momentum or direction. The longitudinal acoustic mode (dashed line) has the highest velocity. Transverse acoustic (dash-dot) and optical modes (dotted line) are slower due to their flat dispersion, especially at higher energies corresponding to phonons near the Brillouin zone boundary.

In Fig. 3.8, we find that phonon absorption is strongly dominated by TA phonons. Curves are presented for electron temperatures ranging from 300 K (bottom curve) to 10,000 K (top curve), corresponding to electric fields from 1 kV/cm to around 300 kV/cm.

Both emission and absorption are in units of inverse seconds. Increasing the applied electric field has the effect of increasing the difference between the electron temperature $T_e$ and the lattice temperature $T_{ph}$. This difference acts to increase both the emission and absorption spectra and favors higher-energy phonons. The largest increases can be noted in LA and optical modes, especially near the Brillouin zone edge, at energies around 48 meV for LA phonons and 65 meV for optical phonons. TA phonons at low electron temperatures are dominated by small-energy phonons, while the opposite is true at high electron temperatures, where the largest increase is also in the zone-edge phonons, at energies around 25 meV. LA phonons have the highest velocities of all phonon polarizations (see Fig. 3.5), so they carry heat energy the fastest. Optical phonons have very short lifetimes and quickly decay into combinations of acoustic phonons, while LA phonons have longer mean free paths, of around 100 nm [22]. As shown in Fig. 3.6, small-energy TA phonons are faster than their zone-edge counterparts, so we can expect an increasing proportion of the heat to be carried by LA phonons at high electron temperatures, corresponding to high electric fields. Similar conclusions can be drawn from the logarithmic plot of the net phonon generation rate, shown in Fig. 3.9. The logarithmic plot shows a nearly linear dependence of the optical phonon generation on phonon energy, indicating an exponential relationship between phonon frequency and generation rate. The most drastic dependence on the electric field strength and electron temperature is again in the LA phonon branch, which increases from very small contributions at low fields to nearly even levels with the optical phonon generation at very high electric field strengths. This is especially true for higher-energy LA and LO phonons with energies near 48 meV, corresponding to short-wavelength phonons near the Brillouin zone edge. These results and trends agree with Monte Carlo calculations of emission spectra in silicon by Pop et al. [4] and similar results for a silicon device by Rowlette and Goodson [22]. Both of these Monte Carlo calculations show that the largest increase in the phonon emission at high fields is in the longitudinal modes around 40 meV.

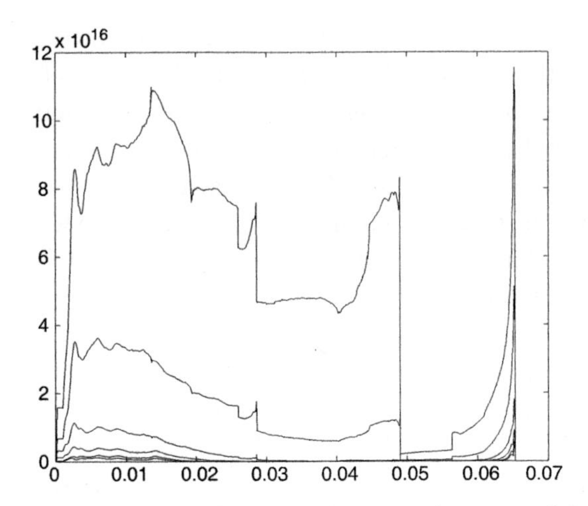


**Figure 3.7** Plot of the rate of phonon absorption (per second) for electron temperatures ranging from 300 K (bottom curve) to 10,000 K (top curve), corresponding to electric fields from 1 kV/cm to around 300 kV/cm. Increase in the applied field causes a difference between $T_e$ and lattice temperature $T_{ph}$ and acts to increase the rate of absorption, especially in the longitudinal acoustic and optical modes.

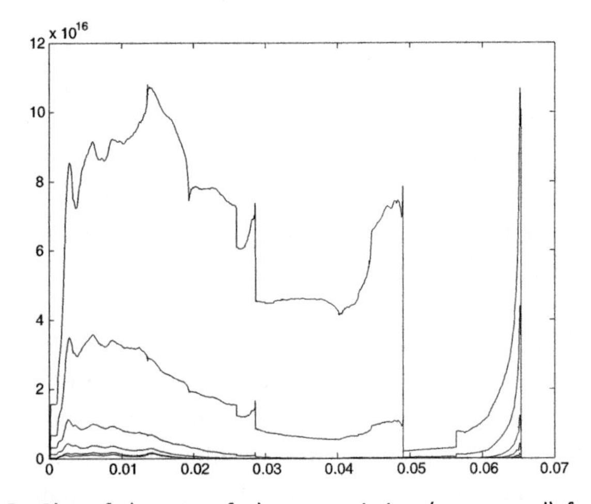


**Figure 3.8** Plot of the rate of phonon emission (per second) for electron temperatures ranging from 300 K (bottom curve) to 10,000 K (top curve), corresponding to electric fields from 1 kV/cm to around 300 kV/cm. The rate of phonon absorption increases with the temperature of electron distribution and favors higher-energy and zone-boundary phonons. At the highest electron temperature, of 10,000 K, we can see sharp increases in the emission of phonons near the highest energy for each mode. This is most prominent in longitudinal acoustic phonons, around 48 meV, and transverse acoustic modes near 25 meV.

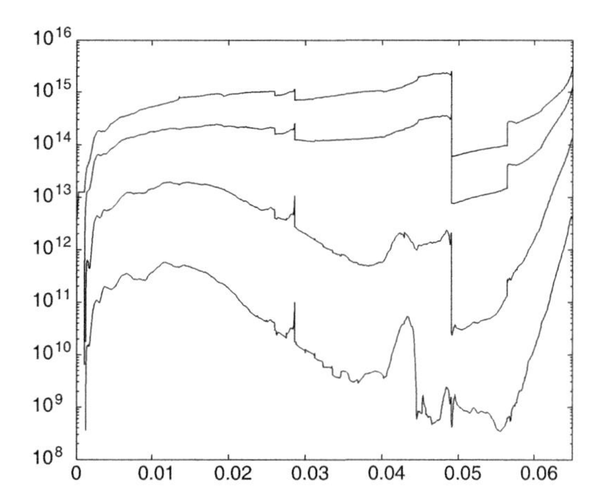


**Figure 3.9** Logarithmic plot of the net rate of phonon generation (per second) for electron temperatures ranging from 300 K (bottom curve) to 10,000 K (top curve), corresponding to electric fields from 1 kV/cm to around 300 kV/cm. This rate is computed from the difference between the total emission and absorption rates and represents the net increase of the phonon distribution function due to coupling between electrons and phonons. The logarithmic plot is linear in the optical region, indicating an exponential energy dependence of emission in the optical branches. This plot also shows that the most dramatic increase with an increasing electric field and electron equivalent temperature is in the longitudinal acoustic branch, especially near the end of the Brillouin zone at energies around 48 meV.

Lastly, we examine the relationship between emission and absorption spectra. Figure 3.10 shows that the ratio of emission and absorption follows a Boltzmann factor $\exp(\omega/k_B T_e)$ at the electron, rather than the lattice, temperature. This ratio is smaller for high electron temperatures and for optical phonons. When electron and phonon temperatures are equal, we have equilibrium and detailed balance holds, producing a ratio equal to the Boltzmann factor. When the electron and phonon populations are out of equilibrium, the electron temperature is higher, producing a net excess of emission over absorption.

## 3.5 Conclusions

We examined the nature of phonon generation in silicon at several strengths of applied electric field. The effect of the field on the electron distribution and equivalent electron temperature was obtained from full-band Monte Carlo simulation for bulk silicon. The electron distributions were used to numerically compute the phonon emission and absorption spectra and learn about their behavior trends at high electron temperatures. The concept of electron temperature was used to understand the relationship between field and heat emission. We identified several trends, in particular that phonon emission at low fields is dominated by low-energy TA phonons. As the temperature of the electron population increases, emission of LA and optical phonons exceeds that of TA modes. It was also found that higher-energy zone-edge phonons increase for all phonon modes at high electron temperatures. Overall, high electron temperatures push the phonon spectra toward higher energies. This favors LA modes the most and shifts the transport of heat more from TA to LA modes. Due to longer lifetimes and focused propagation of LA modes, design optimizations are possible in future devices. These conclusions at high electric fields can be used to enable heat-conscious design of future silicon devices.

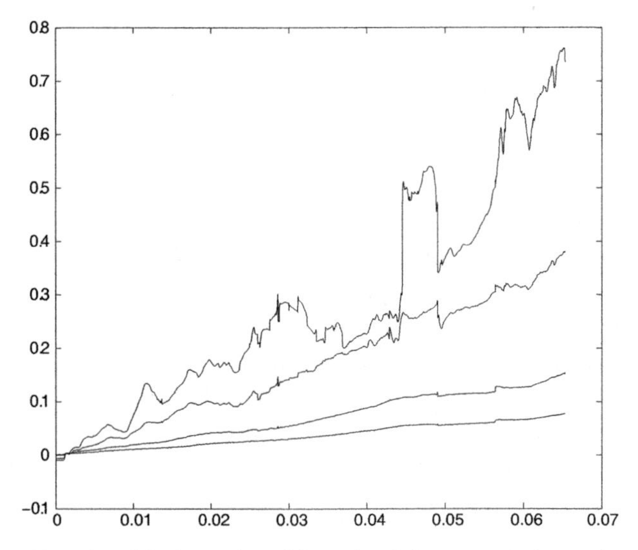


**Figure 3.10** Plot of the logarithm of the ratio of phonon emission to absorption for electron temperatures ranging from 1000 K (bottom curve) to 10,000 K (top curve), corresponding to electric fields from 50 kV/cm to 300 kV/cm. The slope of each curve corresponds to the electron temperature of the distribution from which it was calculated, indicating that the ratio of emission and absorption is given by a Boltzmann factor.

We have developed similar codes for the band structure, phonon dispersion, and computation of phonon emission spectra for strained silicon, as well as most III–V and II–VI compounds of interest. Future work planned includes a comparative study of phonon emission of a range of semiconductor materials to compare their properties from the point of view of thermal emission. This would allow us to draw conclusions about the suitability of use of other materials in future devices.

## References


1. Sinha, S., and Goodson, K. E. (2002). Phonon heat conduction from nanoscale hotspots in semiconductors, in *Heat Transfer 2002: Proceedings of the Twelfth International Heat Transfer Conference*, pp. 573–578.
2. Pop, E., Dutton, R. W., and Goodson, K. E. (2004). Analytic band Monte Carlo model for electron transport in {Si} including acoustic and optical phonon dispersion, *J. Appl. Phys.*, **96**(9), pp. 4998–5005.
3. Fischetti, M. V., and Laux, S. E. (1988). Monte Carlo analysis of electron transport in small semiconductor devices including band-structure and space-charge effects, *Phys. Rev. B*, **38**(14), pp. 9721–9745.
4. Pop, E., Dutton, R. W., and Goodson, K. E. (2005). Monte Carlo simulation of Joule heating in bulk and strained silicon, *Appl. Phys. Lett.*, **86**, pp. 082101–082103.
5. Hess, K. (2000). *Advanced Theory of Semiconductor Devices*, IEEE Press, New York, NY.
6. Weber, W. (1974). New bond-charge model for the lattice dynamics of diamond-type semiconductors, *Phys. Rev. Lett.*, **33**(6), pp. 371–373.
7. Nielsen, O. H., and Weber, W. (1979). Lattice dynamics of group IV semiconductors using an adiabatic bond charge model, *Comp. Phys. Commun.*, **18**, pp. 101–107.
8. Weber, W. (1977). Adiabatic bond charge model for the phonons in diamond, Si, Ge, and α-Sn, *Phys. Rev. B*, **15**, pp. 4789–4803.
9. Devreese, J. T., Doren, V. E. V., and Camp, P. E. V. (eds.) (1981). *Ab initio Calculation of Phonon Spectra*, Plenum Press, New York, NY.
10. Strauch, D., and Dorner, B. (1990). Phonon dispersion in GaAs, *J. Phys.: Condens. Matter*, **2**, pp. 1457–1474.
11. Wolfe, J. P. (1998). *Imaging Phonons*, Cambridge University Press, Cambridge, UK.

12. Ridley, B. K. (1998). *Quantum Processes in Semiconductors*, Clarendon Press, Oxford, UK.
13. Gilat, G. (1972). Analysis of methods for calculating spectral properties in solids, *J. Comput. Phys.*, **10**(3), pp. 432–465.
14. Cohen, M. L., and Bergstresser, T. K. (1966). Band structures and pseudopotential form factors for fourteen semiconductors of the diamond and zinc-blende structures, *Phys. Rev.*, **141**(2), pp. 789–796, doi:10.1103/PhysRev.141.789.
15. Chelikowsky, J. R., and Cohen, M. L. (1976). Nonlocal pseudopotential calculations for the electronic structure of eleven diamond and zinc-blende semiconductors, *Phys. Rev. B*, **14**(2), pp. 556–582, doi:10.1103/PhysRevB.14.556.
16. Sinha, S., Pop, E., Dutton, R. W., and Goodson, K. E. (2006). Non-equilibrium phonon distributions in sub-100 nm silicon transistors, *J. Heat Transfer*, **128**, pp. 638–647.
17. Holland, M. G. (1963). Analysis of lattice thermal conductivity, *Phys. Rev.*, **132**(6), pp. 2461–2471, doi:10.1103/PhysRev.132.2461.
18. Callaway, J. (1959). Model for lattice thermal conductivity at low temperatures, *Phys. Rev.*, **113**(4), pp. 1046–1051, doi:10.1103/PhysRev.113.1046.
19. Allen, B. P. (1983). A tetrahedron method for doubly constrained Brillouin zone integrals application to silicon optic phonon decay, *Physica Status Solidi B*, **120**(2), pp. 529–538.
20. Debernardi, A., Baroni, S., and Molinari, E. (1995). Anharmonic phonon lifetimes in semiconductors from density-functional perturbation theory, *Phys. Rev. Lett.*, **75**(9), pp. 1819–1822, doi:10.1103/PhysRevLett.75.1819.
21. Lai, J., and Majumdar, A. (1996). Concurrent thermal and electrical modeling of sub-micrometer silicon devices, *J. Appl. Phys.*, **79**(9), pp. 7353–7361, doi:10.1063/1.361424, http://link.aip.org/link/?JAP/79/7353/1.
22. Rowlette, J., and Goodson, K. (2008). Fully coupled nonequilibrium electron–phonon transport in nanometer-scale silicon FETs, *IEEE Trans. Electron Devices*, **55**(1), pp. 220–232, doi:10.1109/TED.2007.911043.
23. Conwell, E. M. (1967). *High Field Transport in Semiconductors*, Academic Press, New York, NY.

24. Ridley, B. K. (1998). *Quantum Processes in Semiconductors*, Clarendon Press, Oxford, UK.
25. Gilat, G., and Raubenheimer, L. J. (1966). Accurate numerical method for calculating frequency-distribution functions in solids, *Phys. Rev.*, **144**(2), pp. 390–395, doi:10.1103/PhysRev.144.390.
26. Bir, G. L., and Pikus, G. E. (1974). *Symmetry and Strain-Induced Effects in Semiconductors*, Halsted Press, New York, NY.
27. Ferry, D. K. (2000). *Semiconductor Transport*, Taylor and Francis, New York, NY.
28. Bak, T. A. (1964). *Phonons and Phonon Interactions*, W. A. Benjamin, New York, NY.
29. Kittel, C. (2005). *Introduction to Solid State Physics*, John Wiley and Sons, New York, NY.
30. Duncan, A., Ravaioli, U., and Jakumeit, J. (1998). Full-band Monte Carlo investigation of hot carrier trends in the scaling of metal-oxide-semiconductor field-effect transistors, *IEEE Trans. Electron Devices*, **45**(4), pp. 867–876.
31. Winstead, B., and Ravaioli, U. (2003). A quantum correction based on Schrödinger equation applied to Monte Carlo device simulation, *IEEE Trans. Electron Devices*, **50**(2), pp. 440–446.
32. Tang, J. Y., and Hess, K. (1983). Impact ionization of electrons in silicon (steady state), *J. Appl. Phys.*, **54**(9), pp. 5139–5144.
33. Wingreen, N. S., Stanton, C. J., and Wilkins, J. W. (1986). Electron-electron scattering in nondegenerate semiconductors: driving the anisotropic distribution toward a displaced Maxwellian, *Phys. Rev. Lett.*, **57**(8), pp. 1084–1087, doi:10.1103/PhysRevLett.57.1084.
34. Zebarjadi, M., Bulutay, C., Esfarjani, K., and Shakouri, A. (2007). Monte Carlo simulation of electron transport in degenerate and inhomogeneous semiconductors, *Appl. Phys. Lett.*, **90**(9), 092111, doi:10.1063/1.2709999, http://link.aip.org/link/?APL/90/092111/1.
35. Gilat, G., and Kam, Z. (1969). High-resolution method for calculating spectra of solids, *Phys. Rev. Lett.*, **22**(14), pp. 715–717, doi:10.1103/PhysRevLett.22.715.

# Chapter 4

# Device Simulation, Including the Full Phonon Dispersion

**Zlatan Aksamija**
*University of Massachusetts Amherst, Amherst, MA 01003, USA*
zlatana@engin.umass.edu

## 4.1 Introduction

Joule heating is caused by emission of phonons as electrons traverse through a semiconductor device. In silicon metal-oxide-semiconductor field-effect transistors (MOSFETs), most of the emission is concentrated in the small region where the channel meets the drain. It has been noted in the literature that this causes a hot spot in the device where strong nonequilibrium conditions exist [1]. The emission of phonons has previously been examined [2], but the resulting nonequilibrium temperature conditions in a typical device have not been established to the same level of accuracy. This work aims at using full-band Monte Carlo simulation coupled with phonon transport to quantify the extent and location of Joule heating in a silicon MOSFET. Using these combined approaches, we

*Nanophononics: Thermal Generation, Transport, and Conversion at the Nanoscale*
Edited by Zlatan Aksamija

ISBN 978-981-4774-41-3 (Hardcover), 978-1-315-10822-3 (eBook)
www.panstanford.com

map the propagation of phonons after they are emitted and generate distributions of elevated lattice temperatures. On the basis of this we can explore the effects of phonon heating and propagation in a representative silicon device.

## 4.2 Monte Carlo Device Simulation

Data on scattering events in a device can be obtained from a Monte Carlo simulation. In this work we use a 3D ensemble Monte Carlo simulator with a self-consistent nonlinear Poisson's solver [3] with a quantum correction based on the first moment of the Wigner equation [4]. Our Monte Carlo simulator had, prior to this work, a full-band structure for electrons already implemented [5]. To make the problem tractable, some simplifying assumptions were made. One such assumption often made is that most electrons reside near the bottom of their conduction band and, correspondingly, that most transitions are made between points nearby on the band diagram. This is justified from the perspective of electron-phonon interactions by the fact that the largest phonon energy allowed is around 65 meV, small in comparison to electron energies, which are on the order of an electron-volts. Therefore, we restrict our attention to such transitions. Since the lowest conduction band in silicon is degenerate, it has, in 3D momentum space, six equivalent equal-energy spheres—two on opposite sides of each of the three main directions. The coordinates of these "valleys" are (+/−0.85,0,0), (0,+/−0.85,0), and (0,0,+/−0.85) [6]. These are typically called the $X$ valleys because their directions are all equivalent to the crystallographic $X$ direction. The next valley higher up in energy is the $L$ valley. Therefore, we distinguish between transitions within one $X$ valley and call them intravalley transitions, transitions between two different $X$ valleys, and transitions from an $X$ valley to an $L$ valley, called intervalley transitions. The $X$–$X$ transitions are also split into two types, f and g. Since the transition can occur between a valley and any of the four valleys closest to it or between a valley and another valley opposite to it in momentum space, we distinguish these as f-type (nearest-neighbor transition) and g-type (opposite valley transition) [7]. Graphically, we can see these in Fig. 4.1·

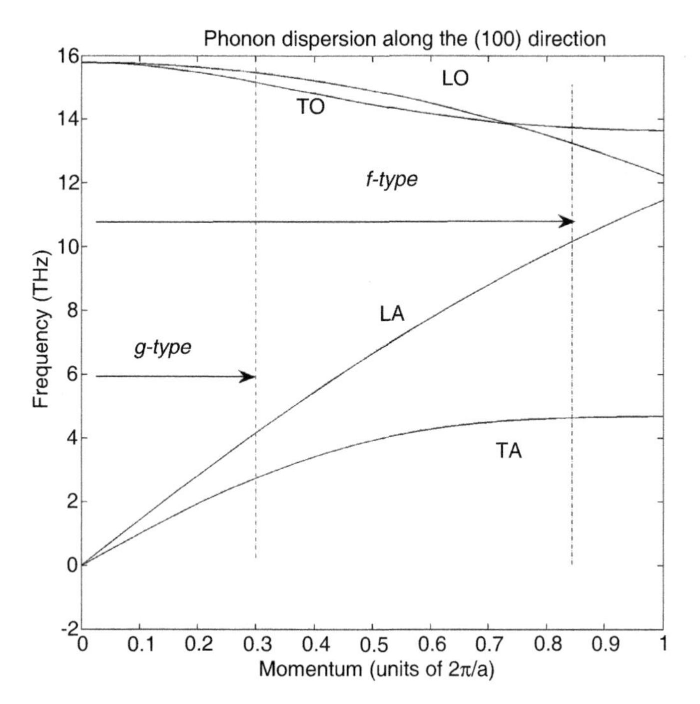


**Figure 4.1** Phonon dispersion along the *X* direction. The reduced momenta for f- and g-type transitions are marked on the graph.

Also in this figure we can note that because we assume electrons are mostly situated at the bottom of each valley, the f- and g-type transitions will have a very limited range of momenta allowed for the phonons generated in those transitions. A g-type transition will involve electrons going from say (0.85,0,0) to (-0.85,0,0). Therefore, the phonon generated will be (1.7,0,0). Reduced to the irreducible wedge (IW) of the first Brillouin zone (FBZ), this is (0.3,0,0). Due to the symmetry of the lattice, we get the same result for all the g-type transitions. This is marked by a dashed line in Fig. 4.1. By a similar line of reasoning, all f-type phonons will be close to the (1,0.15,0.15) point on this graph. This allows us to read off the energies of the phonons involved in each transition and use this to simplify considerably the scattering rate calculations. Using this information we can find scattering probabilities for each of the four different transition types: intravalley, $X - -X$ g-type, $X - -X$ f-type, and $X - -L$ intervalley transitions. Furthermore, for each of these we have three or four phonon branches on which the phonon can be emitted or absorbed, so we end up with a total of 22 different kinds of scattering events. These are tabulated in Table 4.1. Thus, we tabulate transition rates, or probabilities, for only these 22 specific scattering events. The scattering probabilities are also tabulated by the energy of the electron involved in the event and represent the likelihood to scatter by a given mechanism at a given electron energy. They are calculated by integrating Fermi's golden rule over all initial momenta $\mathbf{k}$ with a given energy and all the possible final momenta $\mathbf{k}'$. Since we have made some simplifying assumptions, these rates can now be integrated analytically [8]. There are other scattering types beyond the 22 mentioned here, including surface scattering and ionized impurity scattering, but they are beyond the scope of this work and we will not delve into them any further.

Therefore, whenever a collision happens (and it is decided that a phonon is either emitted or absorbed, the two differing in practice only by a sign), an average phonon energy appropriate for the chosen event type is either added to or subtracted from the current electron energy, according to whether it was an absorption or an emission, respectively. The energy in question is precisely the one listed in Table 4.1 for each event type. Then this final energy of the electron is used to look up the final state of the electron after the collision in a table of electron momenta sorted by energy in the IW of the FBZ. Since the relationship between momentum and energy is bijective, the said relationship can be inverted and thus tabulated for lookup of momentum by energy. This is done on a fine grid and interpolated so that no precision is sacrificed. Thus we end up with a table of momenta that we can simply read off, given our knowledge of the energy after the scattering. Then the actual final state in the complete 3D momentum space is chosen randomly by symmetry considerations, in accordance with the specific type of scattering that occurred [5]. For example, if we have a nearest-neighbor equivalent valley scattering, often termed "*X*–*X* f-type scattering," then we must make sure that the largest component of the momentum in the final state is perpendicular to the largest component of the initial state momentum, while others can have their signs and ordering chosen at random. Thus we end up with a scattering into one of the valleys that is adjacent or closest to the one the electron was in prior to collision, and the remaining components of the momentum are randomizing.

To ensure accuracy of results for phonon events, a full phonon dispersion is included. It is calculated from an adiabatic bond charge model and tabulated for lookup [9]. An iterative algorithm introduced by Pop et al. [6] was adapted in order to make all scattering events involving phonons energy and momentum conserving with the full phonon dispersion relationship. The algorithm starts at each scattering event with an estimate for the energy of the phonon involved. This energy can be calculated from the momenta involving transitions between the bottoms of the $X$ and $L$ equivalent valleys in silicon [8]. Then the final state can be looked up from a table of electron energies and momenta. Finally, this result is checked with the phonon dispersion to ensure energy conservation. If energy is not conserved to within a small tolerance dictated by collision broadening [10], then the final state is rejected and a new state is sought, as shown in Fig. 4.2.

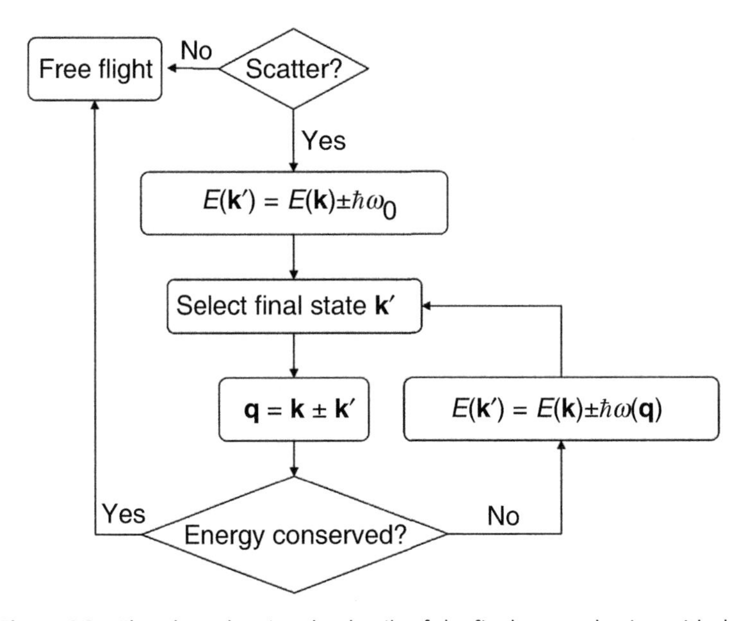


**Figure 4.2** Flowchart showing the details of the final state selection with the full phonon dispersion. This algorithm is applied at each iteration in order to ensure energy conservation with the phonon dispersion, which is tabulated from the adiabatic bond charge algorithm.

**Table 4.1** Scattering events and their phonon energies

| Event | Valley | Sign | Branch | Energy |
|---|---|---|---|---|
| 1 | Acoustic | Absorption | TA/LA | 0 to 45 meV |
| 2 | Intravalley | Emission | TA/LA TA | 0 to 45 meV |
| 3 | *X*–*X* f | Absorption | LA/LO | 18.9582 |
| 4 | | | TO | 47.3954 |
| 5 | | | TA | 59.0288 |
| 6 | *X*–*X* g | | | 12.0643 |
| 7 | | | LA | 18.5273 |
| 8 | | | TO/LO | 62.0449 |
| 9 | *X*–*X* f | Emission | TA | 18.9582 |
| 10 | | | LA/LO | 47.3954 |
| 11 | | | TO | 59.0288 |
| 12 | *X*–*X* g | | TA | 12.0643 |
| 13 | | | LA | 18.5273 |
| 14 | | | LO/TO | 62.0449 |
| 15 | *X*–*L* | Absorption | TO | 57.9085 |
| 16 | | | LO | 54.6340 |
| 17 | | | LA | 41.3632 |
| 18 | | | TA | 16.9762 |
| 19 | *X*–*L* | Emission | TO | 57.9085 |
| 20 | | | LO | 54.6340 |
| 21 | | | LA | 41.3632 |
| 22 | | | TA | 16.9762 |

Therefore, if we call the initial momentum $\mathbf{k}$ and the final momentum $\mathbf{k}'$, the phonon momentum $\mathbf{q}$ is simply $\mathbf{q} = +/-(\mathbf{k}' - \mathbf{k})$ [11]. We know by construction that $E(\mathbf{k}') = E(\mathbf{k}) + / - E_{avg}$, where $E_{avg}$ is the phonon energy from Table 4.1, but there is no guarantee whatsoever that $E(\mathbf{q}) = E_{avg}$. In fact, it should never be true, except on the average, as the name of the variable implies. The idea is then simple: Take the phonon momentum obtained at the end of this calculation, namely $\mathbf{q}$, and its corresponding energy $E(\mathbf{q})$ calculated from the phonon dispersion relationship, and simply use that as the next guess for the electron final energy: $E(\mathbf{k}') = E(\mathbf{k}) +/- E(\mathbf{q})$. Then again look up the final state $\mathbf{k}'$ by the method described above with this final energy. This easily turns into an iteration until the energy conservation is established to within a preset tolerance. To ensure fast convergence, the new state is calculated on the previous estimate of the phonon energy. After several iterations, a final state is found that satisfies both momentum and energy conservation. This gives us a more accurate value for the phonon momentum and energy. Another consequence of this process is ensuring that all possible transitions are well represented, not just those with a fixed predetermined phonon energy.

## 4.3 Thermal Properties of Silicon

Once the simulation run is complete, data on all phonon events that occurred are tabulated. Then phonon velocity is looked up from the dispersion relationship for each phonon and each phonon is allowed to move without scattering until the end of the simulation time frame. Finally, the entire simulation region is divided into small cubes with sides of length of 1 nm and the total phonon energy in each cube is computed. The average equilibrium occupation probability of phonons is given by the Bose–Einstein distribution [11] and includes the temperature dependence, as shown in Eq. 4.1. At low phonon energy, when $\hbar\omega < k_B T$, the Bose–Einstein distribution is well approximated by the equipartition expression in Eq. 4.2. This means the probability of finding a mode with energy $\hbar\omega$ occupied is proportional to temperature and inversely proportional to that energy. Consequently, higher-energy modes, especially those on the optical branches, are much less likely to be occupied in equilibrium until the temperature becomes comparable to the energy $k_B T \approx \hbar\omega$. This observation will extend to most thermal properties.

$$\langle n(\omega)\rangle = \frac{1}{e^{\frac{\hbar\omega}{k_B T}} - 1}. \tag{4.1}$$

$$\langle n(\omega)\rangle \approx \frac{k_B T}{\hbar\omega} - \frac{1}{2}. \tag{4.2}$$

The average number of phonons per unit volume can also be found by summing the equilibrium Bose–Einstein distribution from Eq. 4.1 over all phonon modes $q$ and branches $\mu$, as in Eq. 4.3. A curve relating lattice thermal energy $U(T)$ and temperature $T$ is computed by summing the contributions to the thermal energy from all modes $q$ and branches $\mu$. This is equivalent to integrating the phonon energy $\hbar\omega$ and the equilibrium occupation probability $\langle n(\omega)\rangle$, as given by Eq. 4.1, with the total number of modes near each given value of phonon energy, which is expressed by the lattice vibrational density of states (DOS) $D(\omega)$. Combining Eqs. 4.4 and 4.1 and converting the sum over phonon modes to an integral over phonon energy produces the expression in Eq. 4.4. The heat capacity of the crystal is given by the slope of the $U(T)$ curve, so it can be obtained by differentiating the total thermal energy $U(T)$ with respect to temperature (Eq. 4.5).

$$< N(T) > = \sum_{\mathbf{q},\mu} \langle n(\mathbf{q},\mu)\rangle = \int d\omega D(\omega) \frac{1}{e^{\frac{\hbar\omega}{k_B T}} - 1} \tag{4.3}$$

$$U(T) = \sum_{\mathbf{q},\mu} \hbar\omega(\mathbf{q},\mu)\langle n(\mathbf{q},\mu)\rangle = \int d\omega D(\omega) \frac{\hbar\omega}{e^{\frac{\hbar\omega}{k_B T}} - 1} \tag{4.4}$$

$$C(T) = \frac{\partial U(T)}{\partial T} = k_B \int d\omega D(\omega) \left(\frac{\hbar\omega}{k_B T}\right)^2 \frac{e^{\frac{\hbar\omega}{k_B T}}}{\left(e^{\frac{\hbar\omega}{k_B T}} - 1\right)^2} \tag{4.5}$$

$$D(\omega) = \int \delta\left(\omega - \omega(\mathbf{q},\mu)\right) d\mathbf{q} \tag{4.6}$$

The DOS is defined as the number of modes near a given energy, as expressed in Eq. 4.6. The phonon DOS is calculated numerically from tabulated dispersion data by applying the linear analytic algorithm proposed in Ref. [12] to Eq. 4.6. The results are shown in Fig. 4.3, where we note the dominance of the optical, especially transverse, modes. The Debye approximation $D(\omega) \approx \omega^3$ is plotted for comparison in order to demonstrate the importance of using the full dispersion instead of the simple approximation $\omega = vq$. Once the DOS curve is obtained numerically, the average number of phonons at each energy $\omega$ is given by the product of the phonon distribution function $\langle n(\omega)\rangle$ from Eq. 4.1 and the number of modes at that value of $\omega$ that is given by the DOS. This product is plotted in Fig. 4.4 for temperatures of 77 K, 300 K, and 600 K, showing that optical modes only start to contribute at higher temperatures above room temperature due to the higher energy, and consequently lower probability of occupancy, of optical branches. Integrating these curves over $\omega$ at each value of temperature, as in Eq. 4.3, produces the total number of phonons per unit volume, or the phonon density. The calculated phonon density shown in Fig. 4.5 gives around 60 phonons in each such cube, or per cubic nanometer, at room temperature, so this presents a sufficient number of phonons to compute a reliable average for the energy in each small cube. Lattice thermal energy $U(T)$ can be computed from the DOS by applying numerical quadrature to the expression in Eq. 4.4, and it is shown in Fig. 4.6. The individual contribution of each of the four phonon branches to the lattice thermal energy is shown in Fig. 4.7, where we can see that the strongest contribution is from the doubly degenerate transverse acoustic modes, while longitudinal modes, both acoustic and optical, have a smaller contribution in equilibrium. The longitudinal optical (LO) mode is the lowest, and it takes the least amount of energy to cause a large increase in its temperature. Using these results, the nonequilibrium temperature in each small 1 nm cube in a device can be found by an inverse lookup, both for the overall temperature and for each phonon branch individually. The heat capacity of the silicon crystal is also plotted in Fig. 4.8. The individual contributions of each of the branches to the heat capacity are plotted in Ref. 4.9, showing again the reduced contribution from the nondegenerate longitudinal branches. Since the heat capacity of longitudinal branches is smaller than that of transverse branches, we expect that the temperatures reached by the longitudinal phonons will be higher than those of their transverse counterparts, for both acoustic and optical phonons.

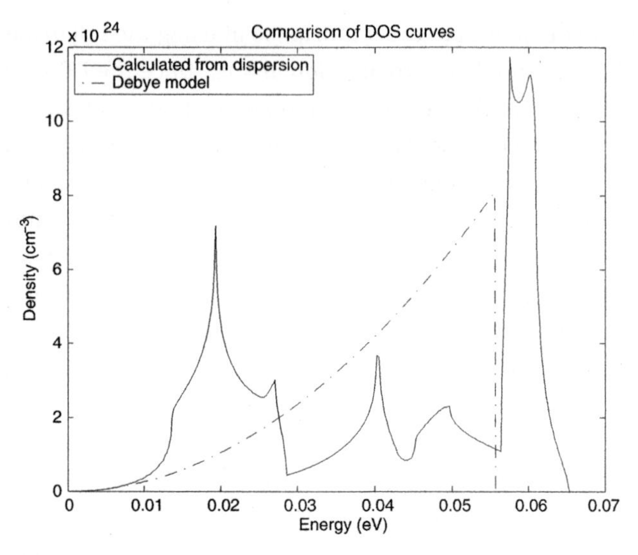


**Figure 4.3** The phonon DOS, computed numerically from the dispersion data using the GR algorithm. The DOS shows that optical modes dominate, especially the transverse modes. The simple analytical cubic Debye approximation is also shown (dashed line), highlighting the significance of considering the full phonon dispersion and its impact on the DOS and related properties.

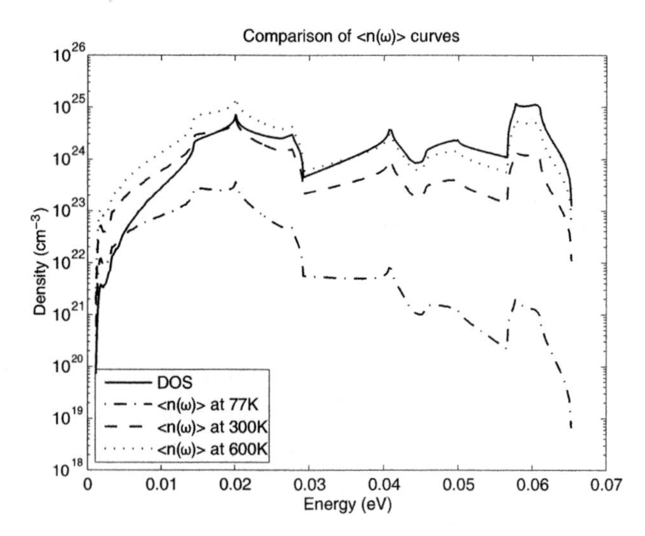


**Figure 4.4** Average number of phonons in equilibrium at temperatures of 77 K, 300 K, and 600 K, showing that optical modes only start to contribute at higher temperatures above room temperature. The DOS (solid line) is plotted for reference to show relative contributions of low- and high-energy modes.

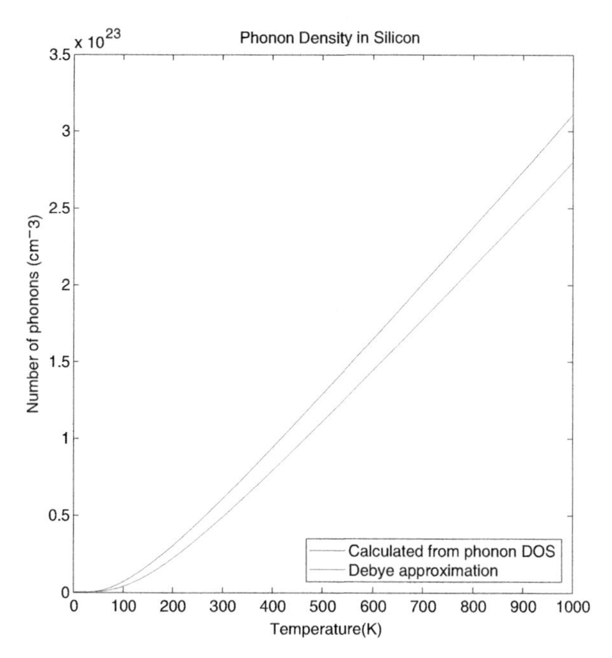


**Figure 4.5** The density of phonons per unit volume computed from the vibrational DOS shows that there are around 60 phonons per cubic nanometer in equilibrium.

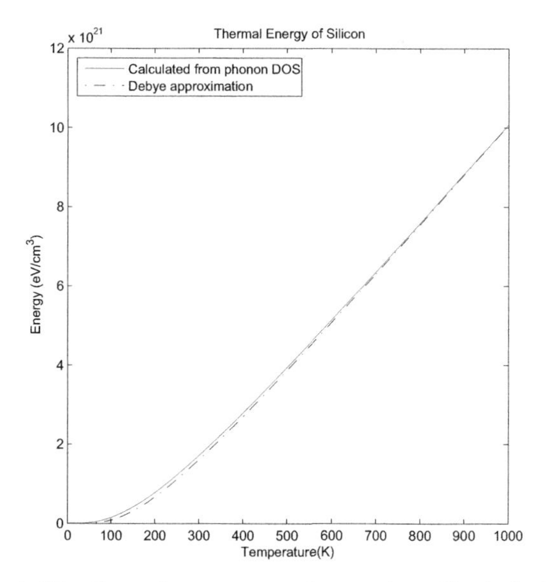


**Figure 4.6** Lattice thermal energy versus temperature. The dashed line is the Debye approximation, which holds well in this case.

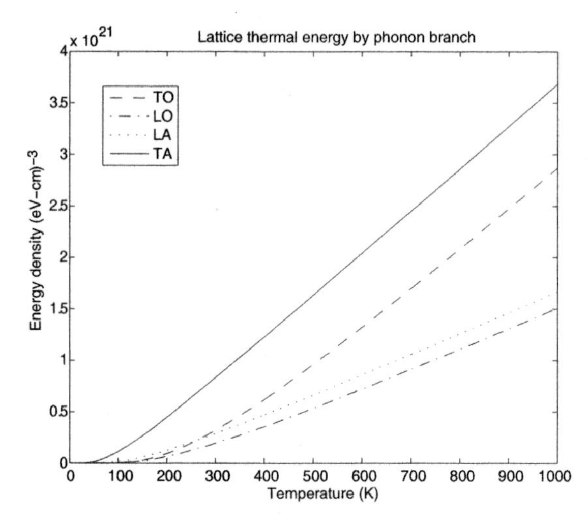


**Figure 4.7** Lattice thermal for each phonon branch. The transverse branches are doubly degenerate, leading to a higher contribution to thermal energy. Optical and longitudinal modes begin to contribute only above room temperature due to their higher vibrational energies. The longitudinal optical mode is the lowest, and it takes the least amount of energy to cause a large increase in its temperature.

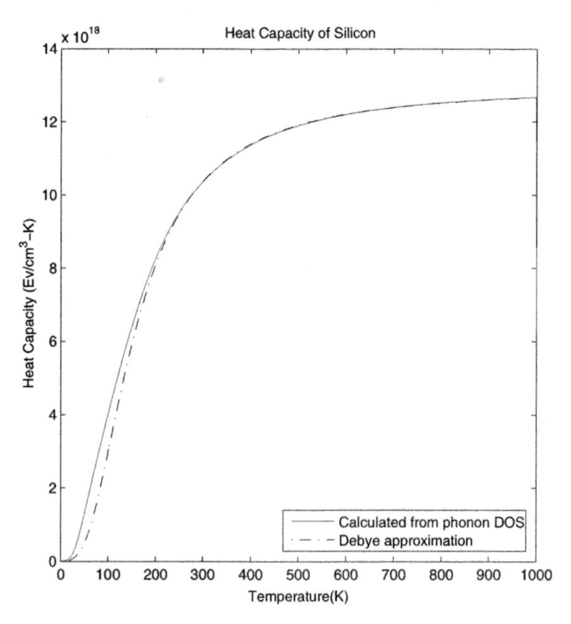


**Figure 4.8** Lattice heat capacity versus temperature. The dashed line is the Debye approximation, which holds well with the exception of low temperatures.

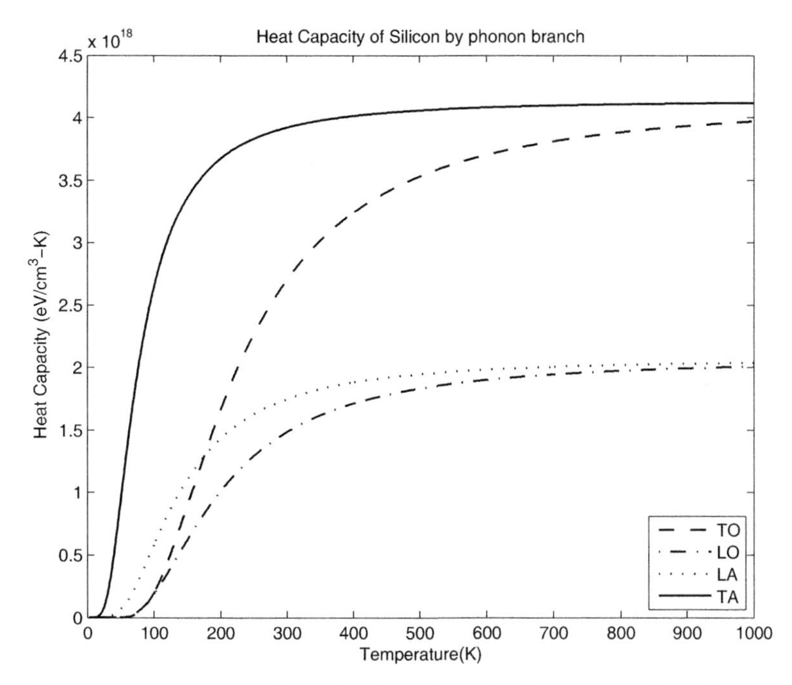


**Figure 4.9** Heat capacity for each phonon branch. The transverse branches are doubly degenerate, leading to a higher heat capacity. The longitudinal acoustic branch becomes flat above room temperature, while optical branches become flat only above 600 K due to their higher vibrational energies. The longitudinal optical has the smallest heat capacity, so it takes the least amount of energy to change its temperature.

## 4.4 Results

The algorithms described in the previous sections were applied to simulate a silicon MOSFET with a channel length of 65 nm. The simulation extended for 100,000 iterations of 0.2 fs for a total duration of 20 ps. The device geometry is shown in Fig. 4.10. It was found that transitions between opposite equivalent valleys dominate the scattering statistics, as shown in Fig. 4.11. Since most electrons reside near the bottoms of the valleys, the momenta of the phonons resulting from equivalent valley transitions fall in very small regions on the optical branches. Because the valleys are located at ⟨0.85 0 0⟩, the momenta of the generated phonons fall very near the point ⟨0.3 0 0⟩. These g-type phonons have reduced momenta, as shown in

Fig. 4.12. The direction of propagation of these phonons is perpendicular to the energy isosurface, shown in Fig. 4.13, which has very flat faces, meaning that phonons propagate in certain preferred directions. Therefore, most of the phonons emitted continue to travel in the same direction as the flow of the electron current. This means very little heat actually flows toward the substrate. We can see this effect from the sequence of snapshots starting at 5 ps in Fig. 4.14, followed by the temperature distribution at 10 ps in Fig. 4.15, 15 ps in Fig. 4.16, and, finally, at the end of simulation, 20 ps in Fig. 4.17. It is also visible in Fig. 4.10 that the cloud of elevated temperature drifts along the channel region.

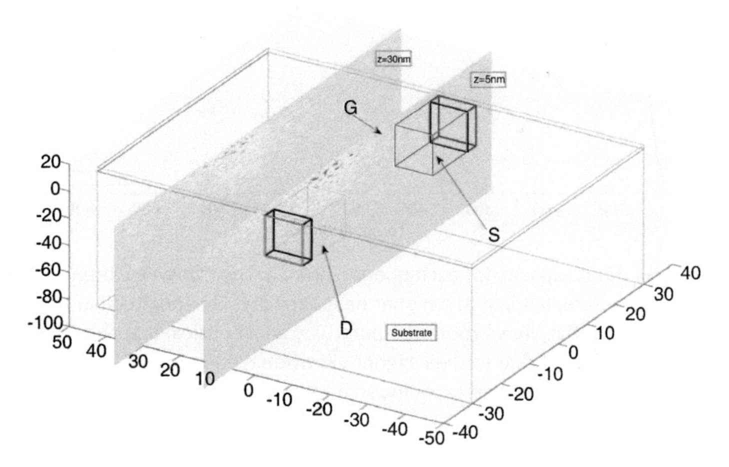


**Figure 4.10** 3D representation of the silicon MOSFET geometry used in this work. The two slices show the cloud of dissipated phonons propagating along the channel. The slices are positioned at distances of 5 and 30 nm from the center of the device.

Figure 4.18 demonstrates that the density of phonons generated during the simulation approaches the phonon DOS and the prominent peak in the optical modes. This result allows us to predict the distribution by energy and branch of phonons created in a realistic device. Furthermore, separating the phonons according to their branches gives a clearer picture of not just where but also how heating occurs. The transverse acoustic branch has the least amount of phonon emission in the simulation, so the temperature map of this branch in Fig. 4.19 shows very little temperature increase. In fact, there is a cooling effect due to the strong absorption of acoustic phonons. The longitudinal acoustic branch participates in phonon generation more strongly, but due to the high group velocity the temperature rise is moderate and spread out widely over the device, as can be seen in Fig. 4.20. The peak temperature is more severe in the optical branches due to their low velocity. This is especially prominent in the transverse optical branch in Fig. 4.21, where, despite the modest amount of heat generated in the transverse optical branches, the peak temperature reaches 35 K above the equilibrium room temperature and remains concentrated in a small hot spot region in the drain, marked by the red outline. The effect is most severe in the LO branch, where the majority of intervalley phonons are generated. Figure 4.22 shows a peak temperature of 500 K, with some motion of optical phonons present due to the low group velocity of LO phonons. These results may have important consequences for future designs that take heat issues into consideration.

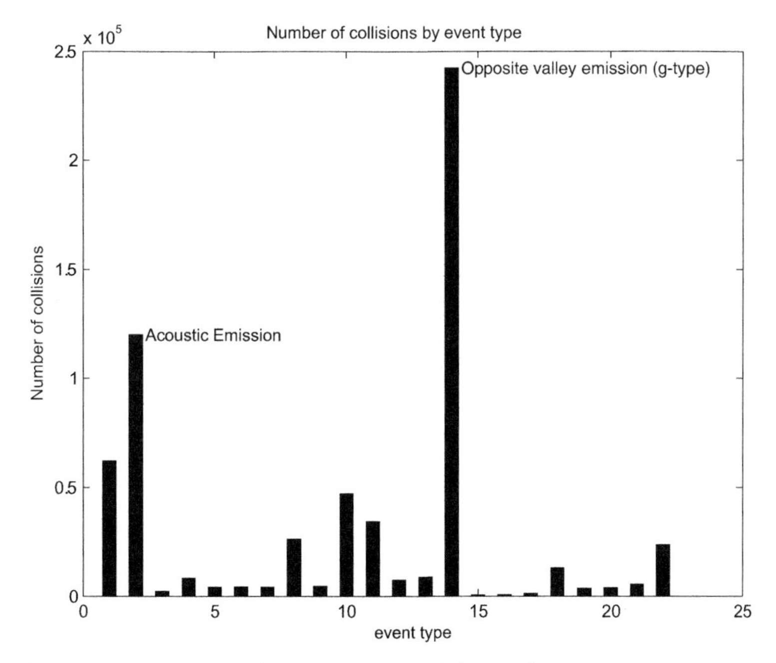


**Figure 4.11** Number of phonons emitted for each event type. Opposite equivalent valley and acoustic intravalley transitions are marked. Opposite equivalent valley transitions dominate.

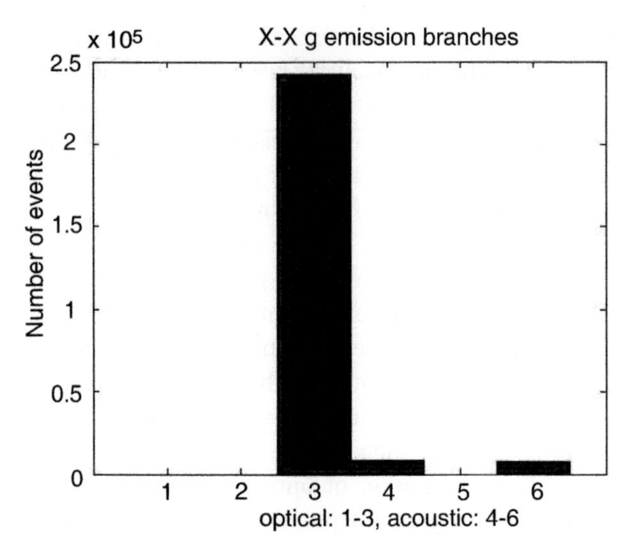


**Figure 4.12** Histogram of phonon branches for opposite equivalent valley phonon emissions. Longitudinal branches dominate this type of transition.

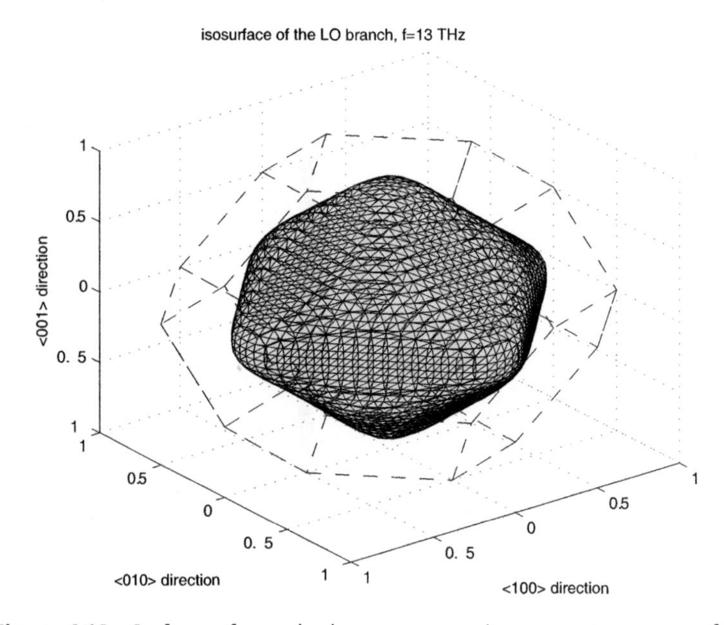


**Figure 4.13** Surface of equal phonon energy in momentum space for the longitudinal optical branch. The direction of phonon propagation is perpendicular to this surface.

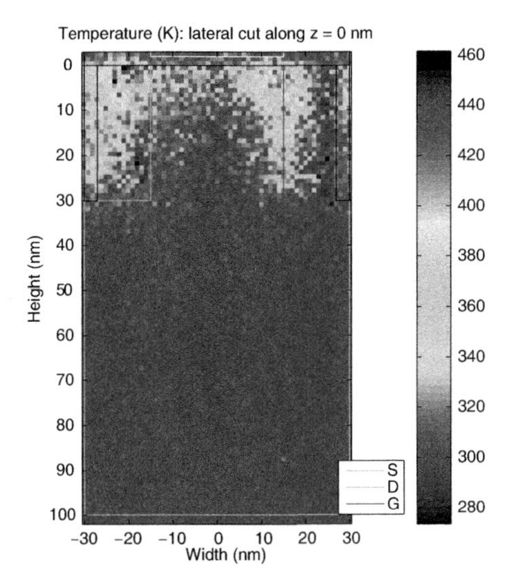


**Figure 4.14** Map of temperature distribution in the 65 nm channel MOSFET device after 5 ps. The initial temperature peak is visible in the drain region.

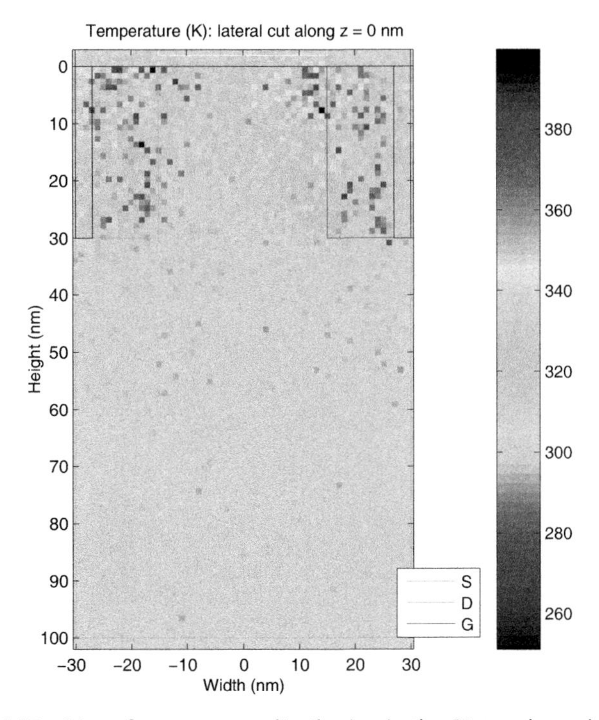


**Figure 4.15** Map of temperature distribution in the 65 nm channel MOSFET device after 10 ps. Phonons begin to move out of the hot spot region.

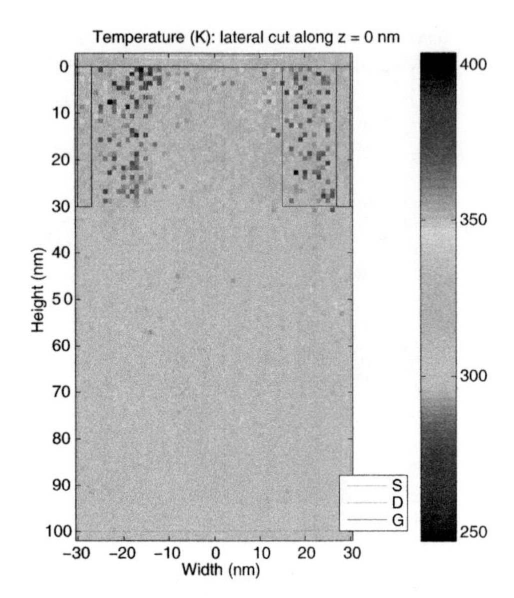


**Figure 4.16** Map of temperature distribution in the 65 nm channel MOSFET device after 15 ps.

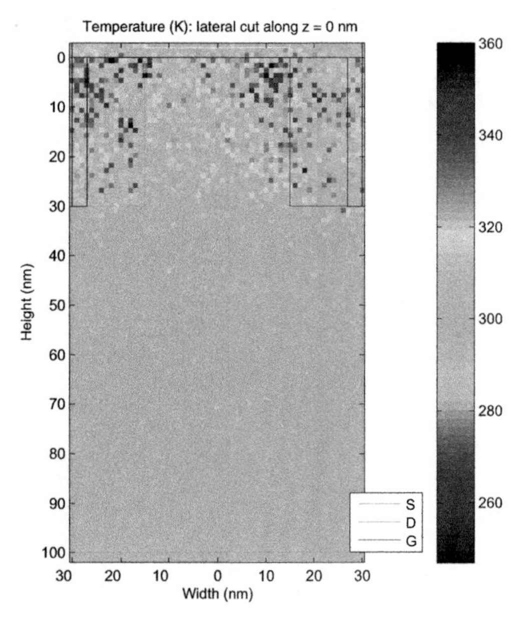


**Figure 4.17** Map of temperature distribution in the 65 nm channel MOSFET device after 20 ps. The clouds of elevated temperature have moved to the left of the source and drain regions.

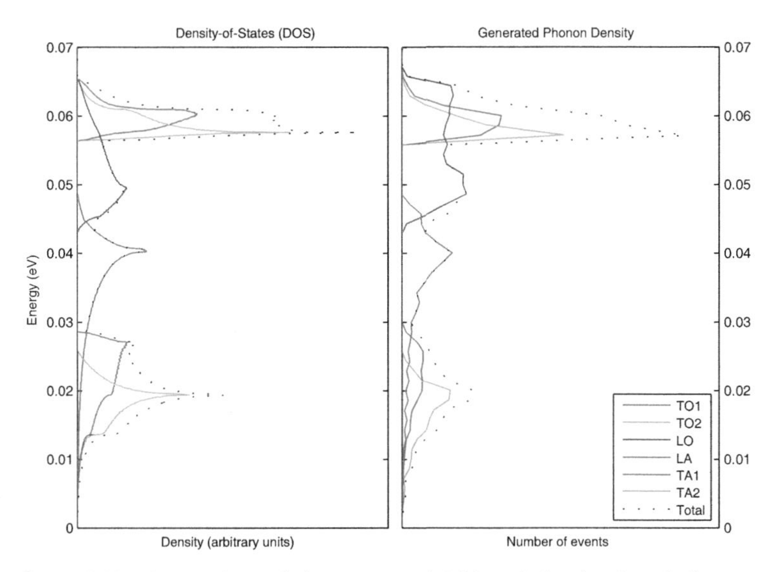


**Figure 4.18** Comparison of the computed DOS and the density of phonons generated during the simulation. The image on the right shows that the density of phonons produced in a silicon MOSFET is well approximated by the phonon DOS.

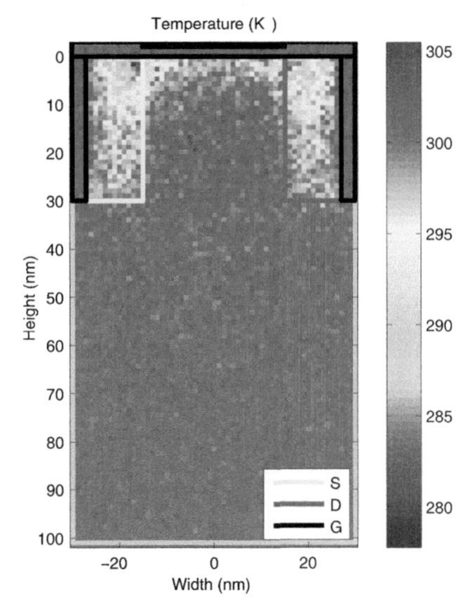


**Figure 4.19** Map of temperature distribution of transverse acoustic phonons in the 65 nm channel MOSFET device after 20 ps.

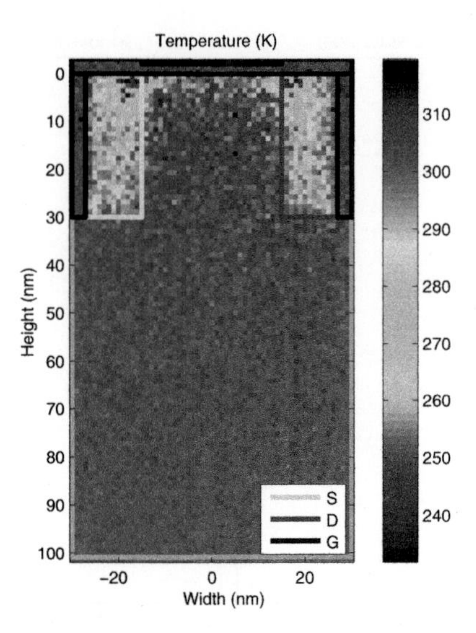


**Figure 4.20** Map of temperature distribution of longitudinal acoustic phonons in the 65 nm channel MOSFET device after 20 ps. Longitudinal acoustic phonons have the highest group velocity, which is evident from the diffused temperature distribution of this branch.

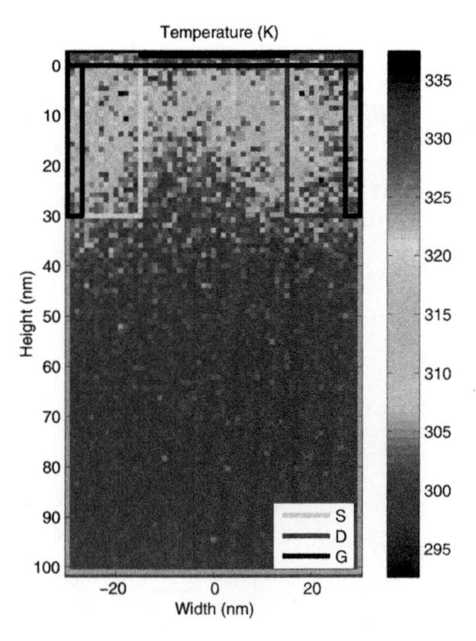


**Figure 4.21** Map of temperature distribution of transverse optical phonons in the 65 nm channel MOSFET device after 20 ps. The low group velocity of transverse optical phonons leads to a small temperature peak in the location at the edge of the drain where most of the phonon emission occurs.

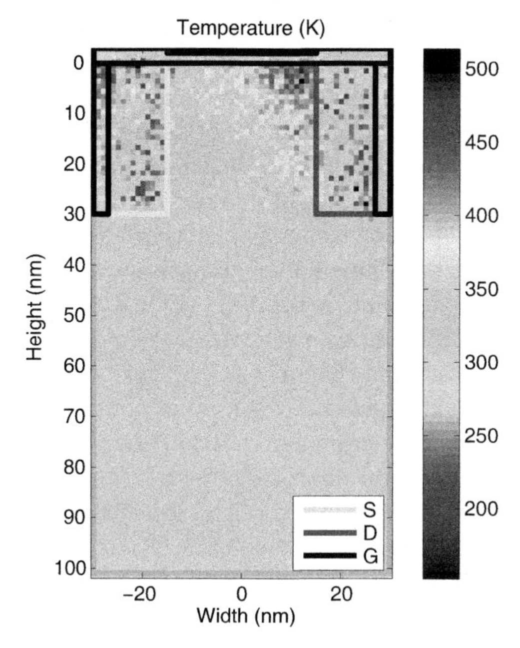


**Figure 4.22** Map of temperature distribution of longitudinal optical phonons in the 65 nm channel MOSFET device after 20 ps. The heating is most severe in this branch, leading to a peak temperature of nearly 500 K.

## 4.5 Conclusions

This work allows us to probe the important process of heat generation in semiconductors and learn about its mechanisms. We can perform accurate calculations of phonon generation in silicon and extract accurate statistics about the process of scattering in silicon, allowing us to learn more about heat in semiconductors and to think about how to solve the pressing issue of reducing heat in very small and densely integrated circuits. The region where most heat is generated is quite narrow and confined to the area where the source and the drain meet the channel. These areas are also the regions where both electron density and energy are high, so the total electron energy is also very large. Scattering rates are energy dependent (if we ignore any anisotropy), and consequently, the areas of high electron energy are also the areas where total dissipation in the form of phonons is also high.

This chapter examined the generation of heat in silicon MOSFETs using self-consistent Monte Carlo device simulation with a full electron band structure and full phonon dispersion computed from the adiabatic bond charge model. We devised an efficient algorithm for the inclusion of full phonon dispersion in order to account for anisotropy and details of heat transport with great accuracy. Including the full phonon dispersion allows an accurate calculation of phonon velocities, but phonon transport was included in the ballistic regime. We computed the DOS and the lattice thermal energy numerically and used them to generate maps of local temperatures in a representative small-channel MOSFET device. Our results show that most heat is dissipated in the form of optical g-type phonons in a small region in the drain and that the heat flows in a preferred direction aligned with the flow of the electron current. We also show that the distribution of generated phonons in energy closely follows the phonon DOS. The strong generation of LO phonons leads to an accumulation of slow phonons, which need to decay into faster acoustic phonons before they can leave the active region of the device [13]. In subsequent chapters we will consider a phonon Monte Carlo code in order to track not just the motion of phonons after they are generated but also their decay into other modes, which will enable a more accurate picture of self-heating and a determination of the role played by the phonon decay bottleneck. To consider this process in more detail, we must include anharmonic decay of phonons.

## References

1. Sinha, S., and Goodson, K. E. (2002). Phonon heat conduction from nanoscale hotspots in semiconductors, in *Heat Transfer 2002: Proceedings of the Twelfth International Heat Transfer Conference*, pp. 573–578.

2. Pop, E., Sinha, S., and Goodson, K. (2006). Heat generation and transport in nanometer-scale transistors, *Proc. IEEE*, **94**, pp. 1587–1601.

3. Winstead, B., and Ravaioli, U. (2003). A quantum correction based on Schrödinger equation applied to Monte Carlo device simulation, *IEEE Trans. Electron Devices*, **50**(2), pp. 440–446.

4. Tsuchiya, H., and Miyoshi, T. (2000). Quantum mechanical Monte Carlo approach to electron transport at heterointerface, *Superlattices Microstruct.*, **27**, pp. 529–532.

5. Hess, K. (1991). *Monte Carlo Device Simulation: Full Band and Beyond*, Kluwer Academic Press, Boston, MA.
6. Pop, E., Dutton, R. W., and Goodson, K. E. (2004). Analytic band Monte Carlo model for electron transport in {Si} including acoustic and optical phonon dispersion, *J. Appl. Phys.*, **96**(9), pp. 4998–5005.
7. Tang, J. Y., and Hess, K. (1983). Impact ionization of electrons in silicon (steady state), *J. Appl. Phys.*, **54**(9), pp. 5139–5144.
8. Jacoboni, C., and Reggiani, L. (1983). The Monte Carlo method for the solution of charge transport in semiconductors with applications to covalent materials, *Rev. Mod. Phys.*, **55**(3), pp. 645–705.
9. Weber, W. (1977). Adiabatic bond charge model for the phonons in diamond, Si, Ge, and α-Sn, *Phys. Rev. B*, **15**, pp. 4789–4803.
10. Kim, K., Mason, B. A., and Hess, K. (1987). Inclusion of collision broadening in semiconductor electron-transport simulations, *Phys. Rev. B*, **36**(12), pp. 6547–6550.
11. Kittel, C. (2005). *Introduction to Solid State Physics*, John Wiley and Sons, New York, NY.
12. Gilat, G., and Raubenheimer, L. J. (1966). Accurate numerical method for calculating frequency-distribution functions in solids, *Phys. Rev.*, **144**(2), pp. 390–395, doi:10.1103/PhysRev.144.390.
13. Rowlette, J., and Goodson, K. (2008). Fully coupled nonequilibrium electron–phonon transport in nanometer-scale silicon fets, *IEEE Trans. Electron Devices*, **55**(1), pp. 220–232, doi:10.1109/TED.2007.911043.

## Chapter 5

# Anharmonic Decay of Nonequilibrium Intervalley Phonons in Silicon

**Zlatan Aksamija**
*University of Massachusetts Amherst, Amherst, MA 01003, USA*
zlatana@engin.umass.edu

## 5.1 Introduction

In this chapter we study phonons produced by the dominant electron transitions between the equivalent $X$ valleys in silicon. We use the Monte Carlo method first to select stochastically the time between phonon collisions and then to select a final-state pair of phonons from the probability distribution for anharmonic decay. Our results show that g-process phonons decay into one near-equilibrium transverse acoustic (TA) phonon and another intermediate longitudinal phonon either on the acoustic or optical branch. This second phonon has energies between 40 and 50 meV and undergoes

*Nanophonics: Thermal Generation, Transport, and Conversion at the Nanoscale*
Edited by Zlatan Aksamija

ISBN 978-981-4774-41-3 (Hardcover), 978-1-315-10822-3 (eBook)
www.panstanford.com

further decay before turning into a pair of near-equilibrium TA phonons, presenting a potential additional bottleneck.

Silicon has been the primary material of complementary metal-oxide-semiconductor (CMOS) technology for decades due to its ability to withstand very high electric fields present in the channels of nanoscale devices. Fields exceeding 100 kV/cm are commonly found in the current deep submicron channel devices. When a high field is applied to a silicon device, such as at the drain end of the channel of a nanoscale metal-oxide-semiconductor field-effect transistor (MOSFET), electrons are accelerated by the field and they reach sufficient energies to begin making transitions between the six equivalent valleys in the crystallographic *X* direction [1]. Such high fields accelerate conduction electrons to energies above 50 meV, sufficient to produce intervalley longitudinal and optical phonons. Due to the indirect bandgap structure of silicon, such transitions occur primarily between the six equivalent *X* valleys, which are located around the point $\langle 0.85\ 0\ 0 \rangle$. Two types of transitions between the valleys are possible. One is a back-scattering type of transition, where an electron transitions to a valley in the direction opposite from the valley from which it originated. This transition is called g-type, and it has, in general, the effect of reversing the electron's momentum vector. The phonons produced by this transition are also referred to as g-type phonons. The other significant intervalley transition, called f-type, is between nearest-neighbor valleys. Both of these transitions are illustrated in Fig. 5.1. Intervalley transitions typically involve long-wavelength optical and longitudinal phonons, and they can also be classified into f-type and g-type phonons according to the type of transition from which these phonons originated. Since the six valleys are located at $\mathbf{k} = \langle 0.85\ 0\ 0 \rangle$, the phonons involved in the intervalley transitions can be found from momentum conservation by taking the difference between the locations of the initial and final valleys of the electrons. In this study, we are interested primarily in the behavior of intervalley phonons after they are generated by the electron transitions, because they are most significant to CMOS technology and are seen as a potential bottleneck to heat transfer [2].

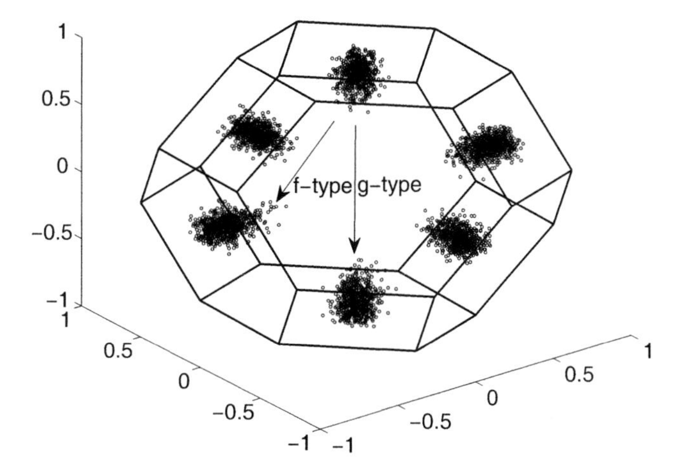


**Figure 5.1** Momenta of electrons in the first Brillouin zone with intervalley transitions between the equivalent *X* valleys marked. Transitions between opposite *X* valleys are g-type, and transitions between neighboring *X* valleys are f-type.

## 5.2 Intervalley Phonon Emission

The momentum of the g-type phonon can be computed from momentum conservation by taking the difference between initial and final electron states. This gives a phonon momentum $\mathbf{q} = \langle 1.7\ 0\ 0 \rangle$, which, when reduced to the first Brillouin zone (FBZ), is $\langle 0.3\ 0\ 0 \rangle$. Similarly, f-type phonons have momenta of $\mathbf{q} = \langle 0.85\ 0.85\ 0 \rangle$, which again reduced to the FBZ become $\mathbf{q} = \langle 1\ 0.15\ 0.15 \rangle$ [1]. On the basis of this observation, we will explore g-type phonons, which have momentum vectors centered around $\mathbf{q} = \langle 0.3\ 0\ 0 \rangle$, as shown in Fig. 5.2, and f-type phonons, which have momenta centered around the point $\langle 1\ 0.15\ 0.15 \rangle$, as shown in Fig. 5.3. Once emitted by the electron-phonon scattering process, intervalley phonons will move slowly due to their low group velocity [3], which is especially prominent in the longitudinal optical (LO) phonon branches, which account for half of the emission in bulk and over 90% of emission in strained silicon [2]. Phonons can decay due to the anharmonicity of the atomic potentials. Anharmonic decay is caused by the cubic

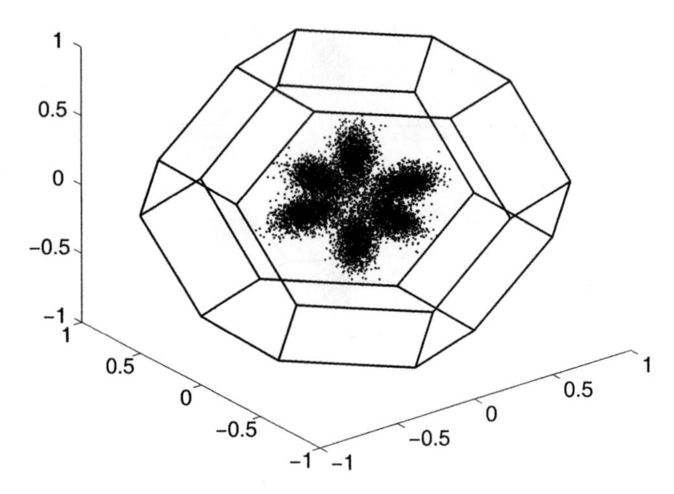


**Figure 5.2** Distribution of phonons produced by g-type electron transitions. The distribution is centered around the point ⟨0.3 0 0⟩ in each direction.

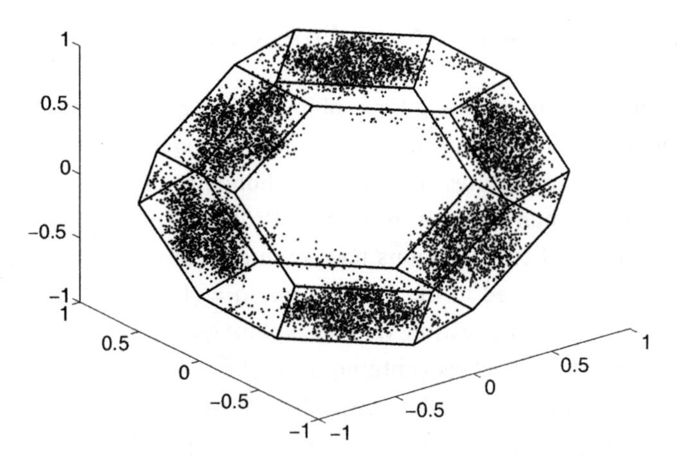


**Figure 5.3** Distribution of phonons produced by f-type electron transitions. The distribution is centered around the point ⟨1 0.15 0.15⟩ in each direction.

and higher terms in the crystal potential. The strongest of these terms is cubic, and it gives rise to the three-phonon process, where a phonon breaks up, or decays, into two phonons, possibly on different branches, while at the same time conserving total crystal momentum and energy [4]. This implies that the strongest decay path is through the three-phonon process, which allows the phonons generated by the electron-phonon coupling to decay into pairs of phonons, each with smaller energy but with higher group velocities [2]. The probability of every such decay will be obtained from perturbation theory, which says that the probability of the three-phonon decay is given by the product of the cubic matrix element and a resonance factor, which ensures energy conservation. The resonance factor, in the limit of infinite collision time, approaches the Dirac delta function, which gives perfect energy conservation. When transitions are so frequent that the average time between collisions, given by the inverse of the transition rate, is small, then the resonance factor is not a perfect Dirac delta but, instead, takes on a distribution around the energy-conserving pair [5]. On the basis of perturbation theory, lifetimes of acoustic phonons were computed in the early works of Klemens [6], Callaway [7], and Holland [8]. Lifetimes of optical phonons were also first considered by Klemens [9] on the basis of a simple model of the lattice consisting of a linear chain of atoms. This assumption produces convenient forms for both the phonon optical and acoustic modes. Klemens also assumed that the optical phonons decay into a pair of acoustic phonons on the same branch but with opposite momenta, which does not apply to the present case. A more detailed study, based on density functional theory (DFT) calculations [10], focused on the decay of the Raman-active zone center optical mode rather than the technologically significant g-type longitudinal optical (gLO) phonon and showed that, in contrast to Klemens's work, optical phonons in silicon decay into pairs of acoustic phonons mostly involving one longitudinal and one transverse branch. Those results were all brought together, tabulated, and summarized in Ref. [11]. In this paper, Rowlette and Goodson show that the decay rate for the intervalley phonons relevant to transport in silicon can be assumed to be independent of phonon energy and momentum and depends only linearly on temperature.

## 5.3 Monte Carlo Simulation of Anharmonic Phonon Decay

The problem of phonon transport modeling, including selecting final-state pairs for the three-phonon decay process, can be approached by adapting the Monte Carlo method [3], which was developed to solve the problem of electron transport in semiconductors [12, 13]. In this work, the Monte Carlo algorithm is used to select the length of free flight for each phonon in the simulation and to select the momenta of the pair of phonons generated by three-phonon interactions. The first of these two steps starts when a phonon is generated by electron scattering. The average lifetime of optical phonons is known from previous studies and measurements [11] and can be estimated to be around 2 ps for optical phonons at room temperature. The probability of a particle to stay in its present state is described by the Poisson's process, so the time of phonon decay can be chosen stochastically by using the Monte Carlo method [14]. This gives the time of phonon decay as $t_{\text{decay}} = -ln(r_1)\tau_{\text{3-phonon}}$, where $\tau_{\text{3-phonon}}$ is the lifetime of the phonon due to three-phonon anharmonic decay and $r_1$ is a random number uniformly distributed on the unit interval. Once the time of the decay process is found, then we must search for a final state. This second step is accomplished using the rejection algorithm on the probability distribution for anharmonic decay, including the matrix element [15]. The probability distribution of anharmonic decay given in Eq. 5.1 was obtained by Klemens [16] from quantum-mechanical perturbation theory as the product of the anharmonic matrix element for the three-phonon process and the time-dependent resonance factor, plotted in Fig. 5.4.

$$P(\mathbf{q},\mathbf{q}') = |\langle \mathbf{q} | H' | \mathbf{q}' \rangle|^2 \, \frac{1-\cos(\Delta\omega t)}{\Delta\omega^2 t}. \qquad (5.1)$$

Here $\Delta\omega = \omega - \omega' - \omega''$ expresses the net energy exchange between the initial and final states. We use this factor from Eq. 5.1 as a probability kernel in the Monte Carlo method to select the final state of one of the two phonons involved in the anharmonic decay, while the second phonon's final state is then chosen on the basis of momentum conservation. The perturbation Hamiltonian due to cubic anharmonicities is given by the following equation:

$$H' = c_3(\mathbf{q},\mathbf{q}',\mathbf{q}'')a^*(\mathbf{q})a^*(\mathbf{q}')a(\mathbf{q}''), \tag{5.2}$$

where the coefficients $a$ and $a^*$are phonon creation and annihilation operators and the anharmonicity coefficient was derived by Klemens [17] from a Gruneisen model, as shown in Eq. 5.3:

$$c_3(\mathbf{q},\mathbf{q}',\mathbf{q}'') = \frac{2\gamma M v^2}{\sqrt{3G}}\mathbf{q}\mathbf{q}'\mathbf{q}''. \tag{5.3}$$

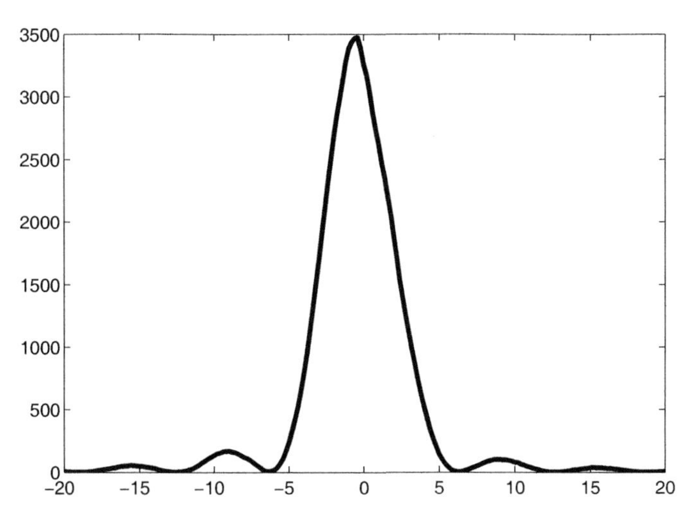


**Figure 5.4** Resonance factor determining energy conservation in the anharmonic phonon decay process. This distribution is given by a sinc function. Most of the distribution is contained within the main lobe between the first zeros occurring when $\Delta\omega \approx \Delta t$.

Placing all the above elements together and expanding the phonon creation and annihilation operators produces the final form of the scattering kernel. The expression (Eq. 5.4) was simplified using the single-mode relaxation theory [18] by assuming that the modes represented by $\mathbf{q}'$ and $\mathbf{q}''$ are near equilibrium.

$$P(\mathbf{q},\mathbf{q}') = 2c_3(\mathbf{q},\mathbf{q}',\mathbf{q}'')^2 \frac{\hbar}{M^3\omega\omega'\omega''}\frac{1-\cos(\Delta\omega t)}{\Delta\omega^2 t}(N'_{\mathrm{eq}} + N''_{\mathrm{eq}} + 1). \tag{5.4}$$

The first of the three phonons involved in the anharmonic three phonon decay is given by the initial phonon and its momentum $\mathbf{q}$. This momentum depends on the type of transition we are interested in (g-type or f-type). Then the second phonon momentum $\mathbf{q}'$ is chosen from a uniform random distribution over the entire FBZ. Finally, the third phonon is produced by momentum conservation as the difference between the initial momentum and the randomly chosen momentum $\mathbf{q}'' = \mathbf{q} - \mathbf{q}' \pm \mathbf{G}$, where $\mathbf{G}$ is a basis vector in reciprocal space, ensuring that the resulting momentum vector $\mathbf{q}''$ falls within the FBZ. For processes where all three vectors are contained in the FBZ, the value of $\mathbf{G}$ is zero. These processes are called Normal. In contrast, those processes where one of the final momentum vectors, in the present case $\mathbf{q}''$, lands outside of the FBZ have nonzero values of $\mathbf{G}$ and are called *Umklapp* [17]. Such processes play a large role in thermal resistance and participate in the decay of intravalley optical phonons. Therefore, both types of processes are considered as candidates for scattering and contribute to the anharmonic decay of gLO phonons. The energy of all three phonons involved is obtained from the full phonon dispersion relationship computed numerically from the adiabatic bond charge model of Weber [19] by diagonalizing the dynamical matrix for each value of the momentum vector [20]. Finally, the total transition probability $P(\mathbf{q}, \mathbf{q}')$ is calculated from Eq. 5.4 and compared with another random number $r_2$ uniformly distributed on the unit interval. If the probability satisfies $P(\mathbf{q}, \mathbf{q}') > r_2$, then the final state pair $\mathbf{q}'$, $\mathbf{q}''$ is accepted; otherwise, the process is repeated. At the end of the procedure, statistics are gathered to plot energies and branches of the selected phonon final states.

## 5.4 Scattering of Acoustic Phonons

For long-wavelength acoustic phonons near the center of the Brillouin zone the energy isosurfaces are close to perfect spheres, so the isotropic approximation can be used. In that case energy $\hbar\omega$ is assumed to depend only on the length of the momentum vector $\mathbf{q} = |\mathbf{q}|$ and not its direction, so a simple quadratic approximation expression were obtained by Pop et al. [1] by fitting experimental data from soft neutron scattering. Such analytical approximations are easier and faster to evaluate and facilitate faster execution of energy lookup for low-energy acoustic phonons. In addition, since the dispersion is almost linear, we can assume that energy is proportional to the length of the momentum vector $\omega = vq$, where $v$ is the speed of sound for each branch (transverse or longitudinal). For this regime, the anharmonic matrix element has an even simpler form given in Eq. 5.5, where the matrix element depends only on the energy and not the momentum of the phonons involved [21]. In addition, the longer lifetimes of acoustic phonons mean that the resonance factor reduces to a perfect Dirac delta distribution, ensuring energy conservation through $\Delta\omega = 0$. This also implies that the decay process will not be very strong for low-energy acoustic modes due to their low density of states in the acoustic limit. Instead, the inverse process occurs where a phonon combines with another low-energy acoustic phonon either on the transverse or on the longitudinal branch and produces a single phonon that is higher in energy [22]. This leads to momentum and energy conservation being expressed as $\mathbf{q}'' + \mathbf{q} + \mathbf{q}'$ and $\omega'' = \omega + \omega'$. These considerations lead to the final form of the coupling probability shown in Eq. 5.6 [18].

$$c_3(\mathbf{q},\mathbf{q}',\mathbf{q}'') = \frac{2\gamma M}{\sqrt{3G}v}\omega\omega'\omega''. \tag{5.5}$$

$$P(\mathbf{q},\mathbf{q}') = \frac{\hbar\omega\omega'\omega''}{2\pi M^3 v}\delta(\omega + \omega' - \omega)(N'_{\text{eq}} - N''_{\text{eq}}). \tag{5.6}$$

To use the rejection method to find the final states after scattering, we rewrite Eq. refeq:acoustic in terms of the energy of the first phonon involved. Then the energy of the first phonon involved is selected by the rejection method on the probability given in Eq. 5.7.

$$P(\omega') = \omega\omega'(\omega + \omega')\left(N_{\text{eq}}(\omega') - N_{\text{eq}}(\omega + \omega')\right). \tag{5.7}$$

Then the energy of the second phonon is found from energy conservation as $\omega'' = \omega + \omega'$, and both their momenta $\mathbf{q}'$ and $\mathbf{q}''$ are then selected on the basis of momentum conservation. Although the emission of optical and large-energy longitudinal phonons is the dominant relaxation mechanism, inclusion of the interactions of acoustic modes is also important for heat transfer as they carry the majority of thermal energy in the quasi-equilibrium regime.

## 5.5 Results and Discussion

Results in Fig. 5.5 show that most of the gLO phonons decay predominantly into one longitudinal, either longitudinal acoustic (LA) or LO, phonon and one TA phonon, in agreement with DFT [10] and molecular dynamics (MD) calculations [4]. The energy distribution of the produced phonons, shown in Fig. 5.6, shows that the LA phonons produced by anharmonic decay are nonequilibrium and have energies between 40 and 50 meV. The equilibrium energy distribution, plotted with a dashed line in Fig. 5.6, shows that most of the equilibrium thermal energy is contained in the TA modes, with a peak around 20 meV. Energy distributions that follow this trend are near equilibrium, while those that have peaks in other branches and at higher energies can be said to be nonequilibrium. Figure 5.7 shows that the phonons produced by the anharmonic decay of gLO phonons furthermore decay themselves into pairs of near-equilibrium TA phonons. The resulting energy distribution after the second decay process, shown in Fig. 5.8, strongly resembles the equilibrium distribution.

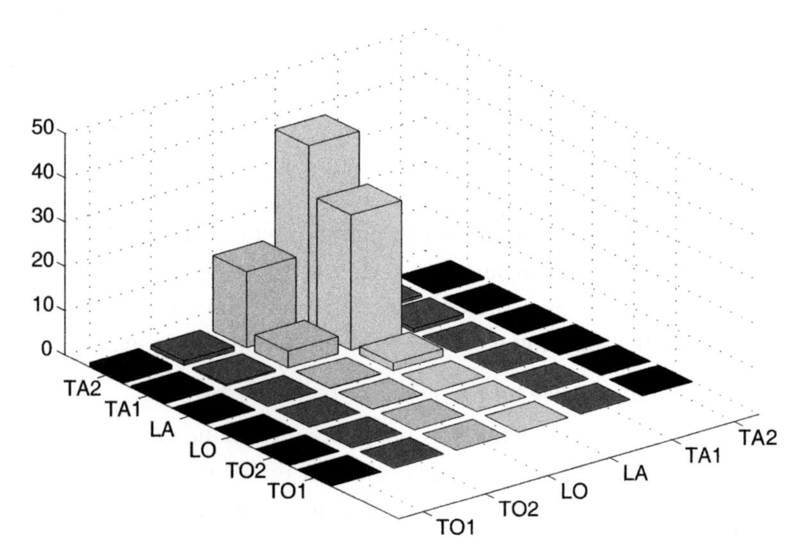


**Figure 5.5** Bar plot of the percentage of phonons on each of the six phonon branches that are produced by the anharmonic decay of g-type LO phonons. Most gLO phonons decay into combinations of TA+LA and some TA+LO phonons, while very few have both phonons on the same branch.

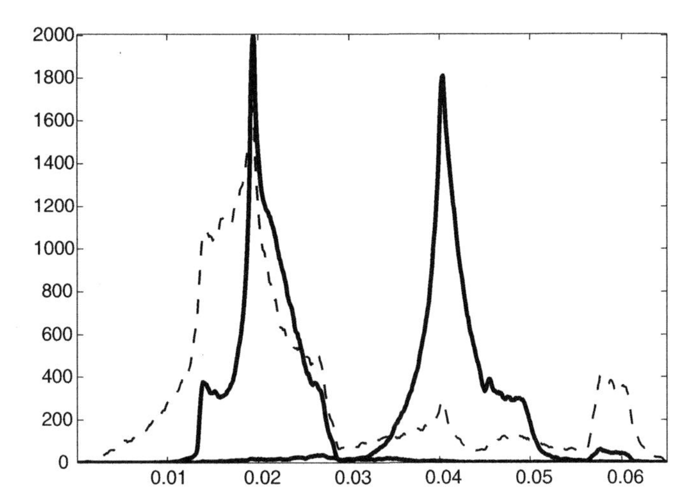


**Figure 5.6** Energy distribution of phonons produced by the anharmonic decay of g-type longitudinal optical (gLO) phonons. Most of the 100,000 simulated phonons break up into one low-energy acoustic and one higher-energy longitudinal phonon. The equilibrium phonon distribution (dashed line) shows that only the lower-energy (below 25 meV) phonon belonging to the TA branch is near equilibrium.

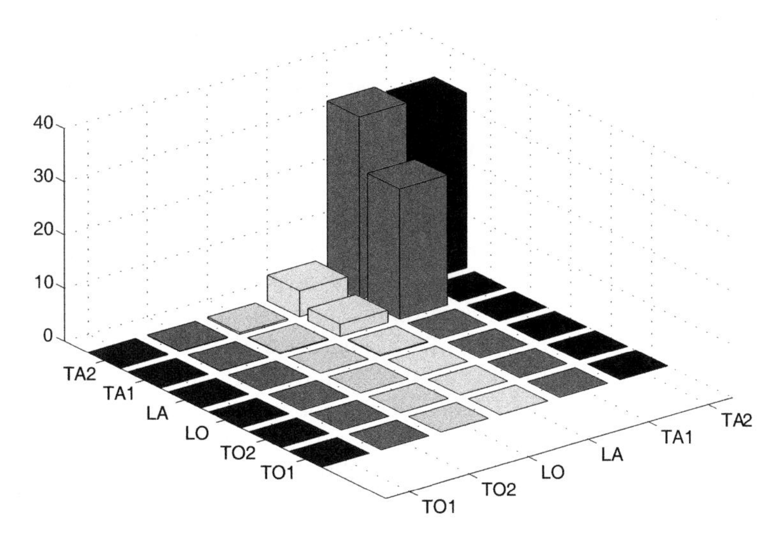


**Figure 5.7** Bar plot of the percentage of phonons on each of the six phonon branches that are produced by the second anharmonic decay of gLO phonons. Most gLO phonons decay into combinations of one TA phonon and another that is on the LA or LO branch and undertakes another decay before it converts into a pair of near-equilibrium TA phonons with energies below 25 meV.

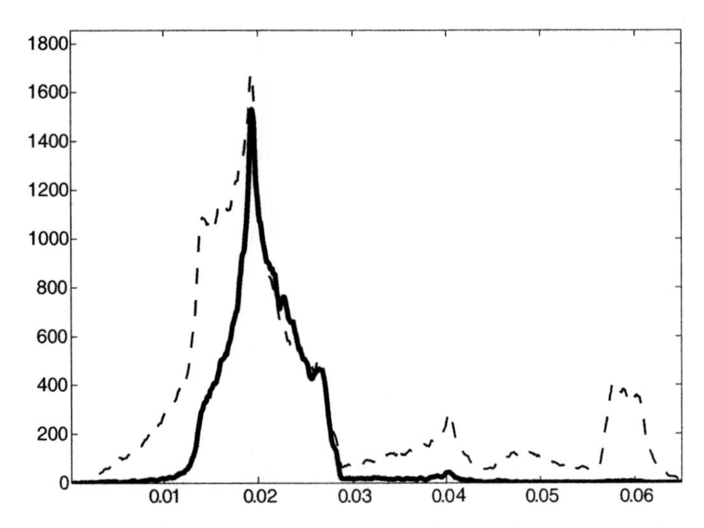


**Figure 5.8** Energies of phonons produced after gLO phonons undergo a second decay, showing that the second anharmonic decay process returns the phonons to equilibrium (dashed line). This implies that slow intervalley optical phonons require additional time to fully decay back to faster-equilibrium TA phonons, posing another potential bottleneck to heat transfer away from the hot spot region in the drain of MOSFET devices.

The other significant optical g-process phonon is the transverse optical branch. Figure 5.9 demonstrates that the g-type transverse optical (gTO) phonon also decays into LA+TA pairs, with very few decays occurring through the Klemens channel where both phonons are of the same polarization. The energy distribution after the gTO decay, shown in Fig. 5.10, also follows a trend similar to gLO and has a near-equilibrium peak in the TA branch and a nonequilibrium component on the LA branch. Several peaks can be noted here, each coinciding with a peak in the phonon density of states (DOS), especially peaks in the two TA branches of around 20 meV and the one in the LA branch of around 40 meV. Both optical g-process phonons have a strong *Umklapp* component accounting for nearly half of the decays, as shown in Fig. 5.11. Consequently, relaxation of g-process phonons toward equilibrium occurs one branch at a time, and the longitudinal branches serve as intermediate branches between the LO and the TA. This can lead to reabsorption of nonequilibrium phonons by the electrons, especially the LA and LO branches, that couple strongly with electrons. Increasing the absorption of longitudinal phonons imparts additional energy on the electron population and can cause more electrons in the tail of the distribution, producing an increase in the number of hot electrons. This effect can have a large impact on the hot electron effect and reliability.

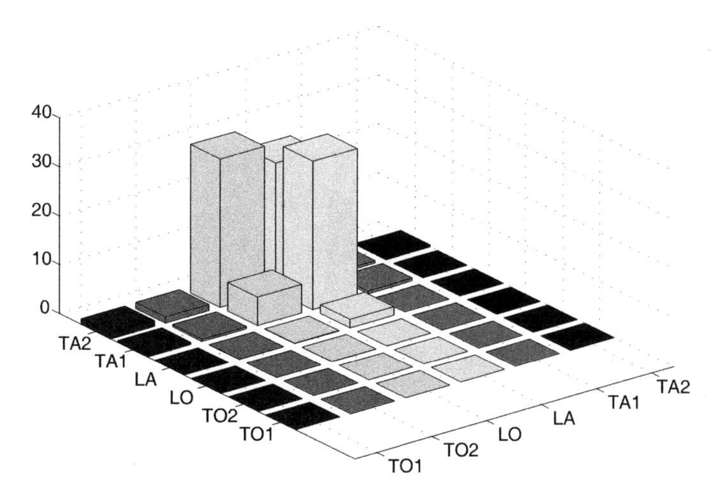


**Figure 5.9** Bar plot of the percentage of phonons on each of the six phonon branches that are produced by the anharmonic decay of g-type TO phonons. Most gTO phonons decay into combinations of TA+LA and also TA+LO phonons, leading to an increase in both LO and LA phonons.

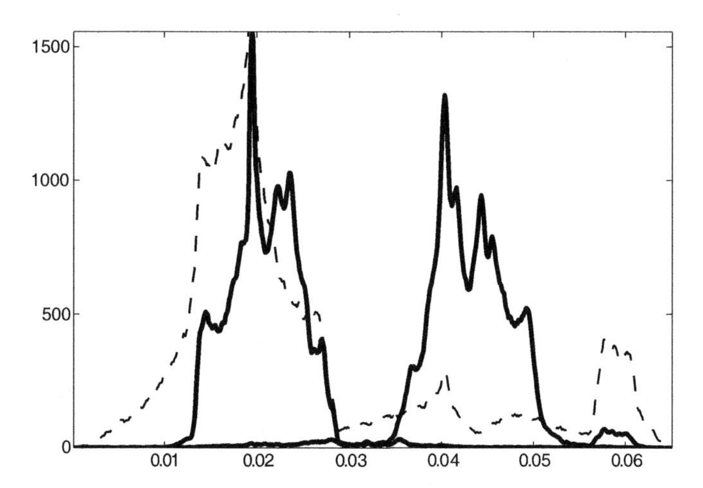


**Figure 5.10** Energies of phonons produced by the anharmonic decay of gTO phonons. The dashed line represents the equilibrium distribution of phonons at room temperature, showing that equilibrium phonons are primarily in the TA branch with energies below 25 meV.

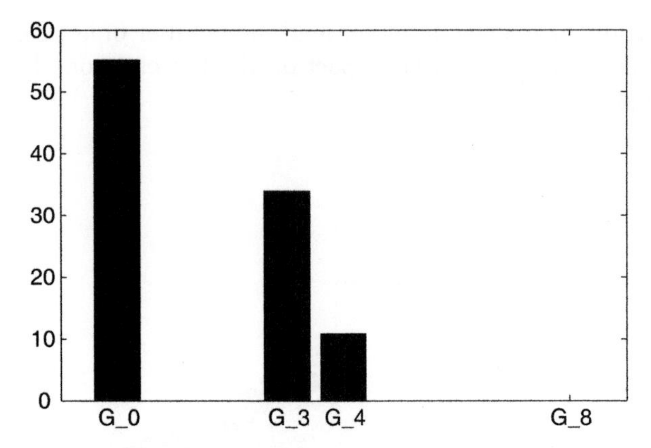


**Figure 5.11** Bar plot of the reciprocal lattice basis vectors **G** involved in the decay process of gLO phonons, showing that close to one half of the decays have a nonzero value and so they are of the *Umklapp* type.

In contrast, the decay of f-process phonons is through the Klemens channel involving two phonons on the TA branch, as shown in Fig. 5.12 for the LO phonon. The energy distribution of the two TA phonons after the decay of f-type longitudinal optical (fLO) phonons in Fig. 5.13 shows both peaks are below 30 meV and coincide with the equilibrium distribution. There is also a significant acoustic component to the intervalley transitions, primarily on the LA branch. Figure 5.14 shows that most of the f-type longitudinal acoustic (fLA) phonons also follow a similar trend and decay through the Klemens channel into a pair of TA phonons. The distribution of energies after the decay of fLA phonons in Fig. 5.15 strongly resembles that of the fLO phonons. Both of the f-process phonons decay predominantly through the *Umklapp* process involving reciprocal basis vectors of magnitudes 3 and 4, as can be seen in Fig. 5.16. Therefore, the f-process phonons are able to decay more efficiently toward equilibrium as they do not involve the LA or LO branch as an intermediate step but, instead, decay directly into pairs of TA phonons that have higher group velocity and heat capacity, leading to a smaller increase in the resulting temperature.

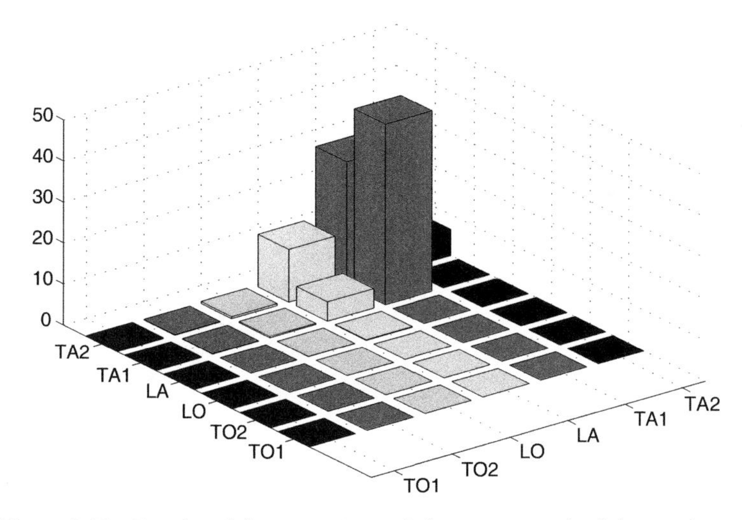


**Figure 5.12** Bar plot of the percentage of phonons on each of the six phonon branches that are produced by the anharmonic decay of f-type LO phonons. Most fLO phonons decay into combinations of TA+TA phonons. This type of decay is called the Klemens channel. The resulting TA phonons are near equilibrium.

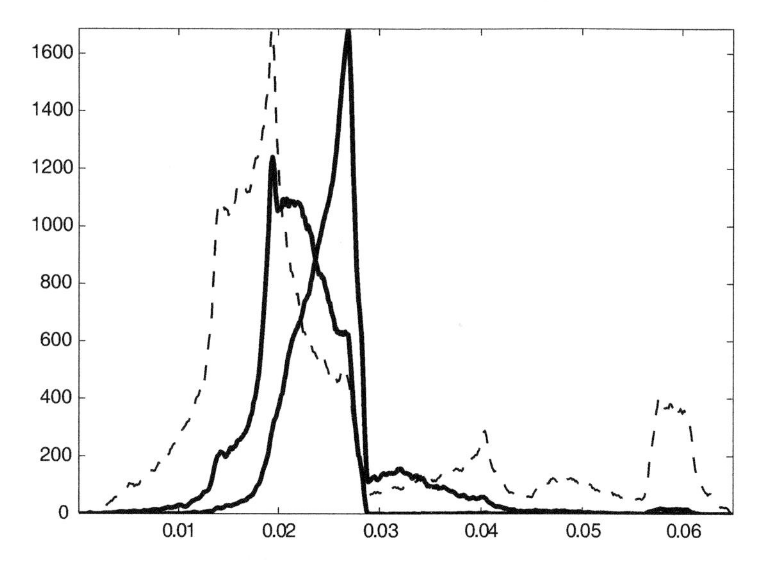


**Figure 5.13** Energies of phonons produced by the anharmonic decay of fLO phonons. Unlike their gLO counterparts, fLO phonons decay primarily through the Klemens channel into pairs of phonons on the same polarization, mainly on the TA branches, meaning that there is less of a bottleneck for f-process optical phonons.

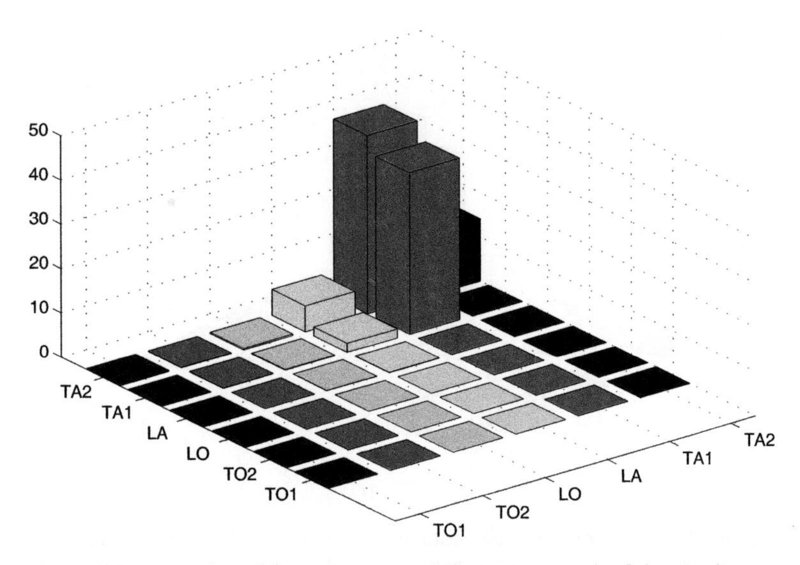


**Figure 5.14** Bar plot of the percentage of phonons on each of the six phonon branches that are produced by the anharmonic decay of f-type LA phonons. Most fLA phonons also decay into combinations of two near-equilibrium TA phonons, presenting less of a bottleneck to phonon transport.

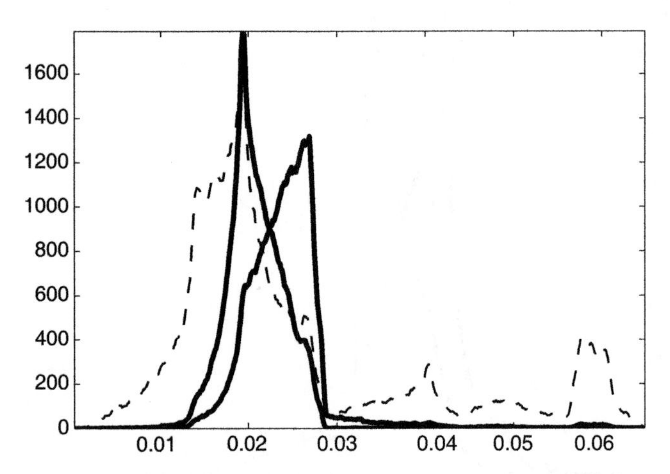


**Figure 5.15** Energies of phonons produced by the anharmonic decay of fLA phonons. Most fLA phonons decay into pairs of TA phonons with energies below 30 meV.

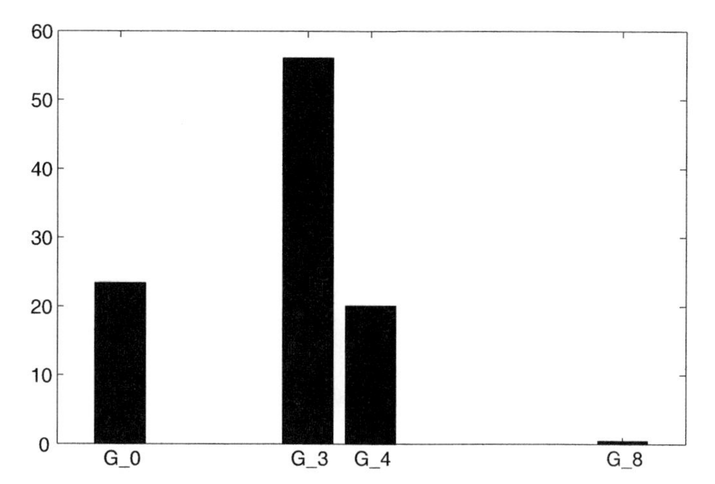


**Figure 5.16** Histogram plot of the reciprocal lattice basis vectors **G** involved in the decay process of f-type phonons, showing that much more than one half of the decays have a nonzero value, so they are of the *Umklapp* type, while very few are of the *Normal* type.

This methodology was coupled to the electron Monte Carlo simulation described in the previous chapter and applied to the simulation of a bulk silicon MOSFET with a 50 nm channel length, shown in Fig. 5.17. The simulation results for the energies of phonons emitted and absorbed are shown in Fig. 5.18. The bar plot of the resulting phonons according to branch and mechanism, shown in Fig. 5.19, has a strong presence of both emission and absorption of LA phonons, as well as a large number of LO phonons, especially for g-process emission. The plot of the temperature distribution along a cut down the center of the 50 nm channel MOSFET device is shown in Fig. 5.20. There is a hot spot in the drain region with a peak temperature of 345 K, or 45 K above the background room temperature. This peak is higher than what would be obtained from heat diffusion alone but much lower than the temperature distribution obtained without considering anharmonic phonon decay. After the decay of optical and LA phonons, heat transfer is nearly diffusive and can be described well by the classical heat equation.

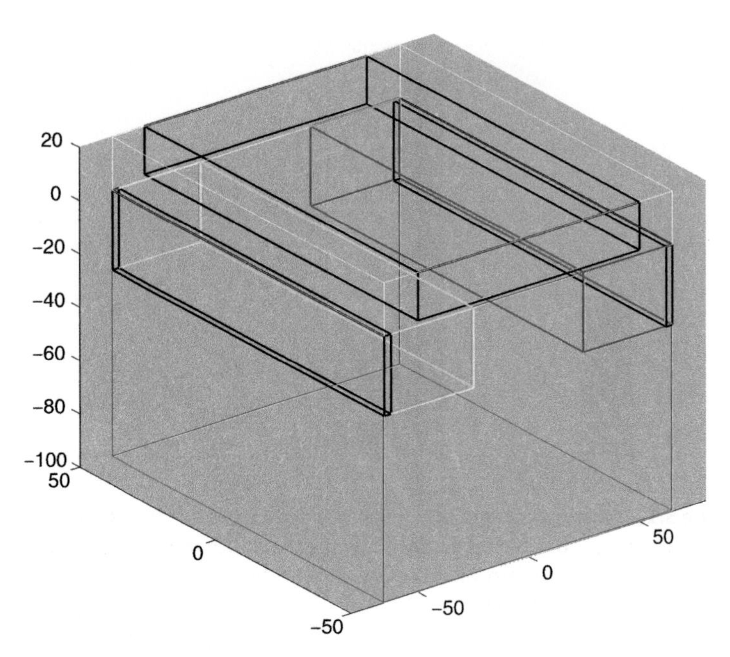


**Figure 5.17** Geometry of the 50 nm channel silicon MOSFET device used in the simulation, showing the position of the source (green), drain (red), and gate regions and their dimensions. Metallic contact regions are outlined in black.

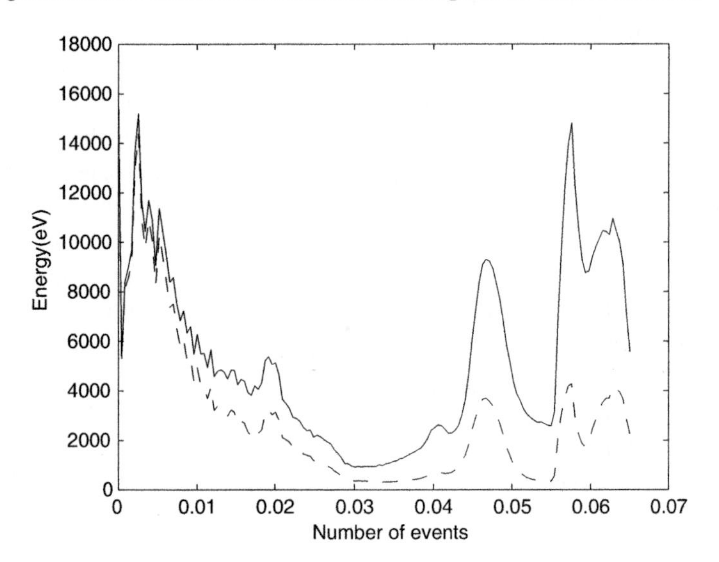


**Figure 5.18** Plot of the energy distribution of phonons emitted and absorbed in the simulation of the 50 nm channel MOSFET device, showing a peak in the longitudinal optical phonons.

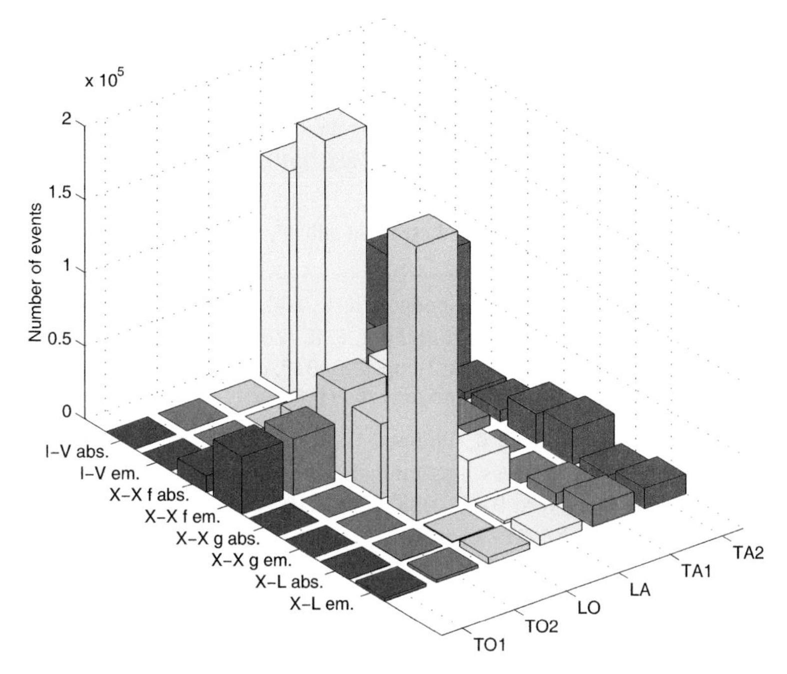


**Figure 5.19** Plot of the mechanisms of phonon emission and absorption in the simulation of the 50 nm channel MOSFET device. There is a strong presence of both emission and absorption of longitudinal acoustic phonons, as well as a large number of longitudinal optical phonons, especially for g-process emission.

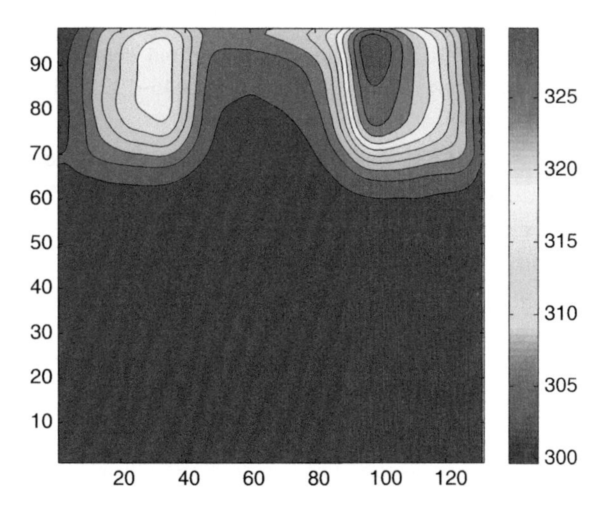


**Figure 5.20** Plot of the temperature distribution along a cut down the center of the 50 nm channel MOSFET device, showing a hot spot in the drain region and a peak temperature of 45 K above room temperature.

## References


1. Pop, E., Dutton, R. W., and Goodson, K. E. (2004). Analytic band Monte Carlo model for electron transport in {Si} including acoustic and optical phonon dispersion, *J. Appl. Phys.*, **96**(9), pp. 4998–5005.
2. Pop, E., Dutton, R. W., and Goodson, K. E. (2005). Monte Carlo simulation of Joule heating in bulk and strained silicon, *Appl. Phys. Lett.*, **86**, pp. 082101–082103.
3. Cahill, D. G., Ford, W. K., Goodson, K. E., Mahan, G. D., Majumdar, A., Maris, H. J., Merlin, R., and Phillpot, S. R. (2003). Nanoscale thermal transport, *J. Appl. Phys.*, **93**(2), pp. 793–818, doi:10.1063/1.1524305, http://link.aip.org/link/?JAP/93/793/1.
4. Sinha, S., Schelling, P. K., Phillpot, S. R., and Goodson, K. E. (2005). Scattering of g-process longitudinal optical phonons at hotspots in silicon, *J. Appl. Phys.*, **97**(2), 023702, doi:10.1063/1.1831549, http://link.aip.org/link/?JAP/97/023702/1.
5. Aksamija, Z., and Ravaioli, U. (2009). Energy conservation in collision broadening over a sequence of scattering events in semiclassical Monte Carlo simulation, *J. Appl. Phys.*, **105**(8), p. 083722, doi:10.1063/1.3116544, http://link.aip.org/link/?JAP/105/083722/1.
6. Klemens, P. G. (1951). The thermal conductivity of dielectric solids at low temperatures (theoretical), *Proc. R. Soc. Lond. A, Math. Phys. Sci.*, **208**(1092), pp. 108–133, http://www.jstor.org/stable/98761.
7. Callaway, J. (1959). Model for lattice thermal conductivity at low temperatures, *Phys. Rev.*, **113**(4), pp. 1046–1051, doi:10.1103/PhysRev.113.1046.
8. Holland, M. G. (1963). Analysis of lattice thermal conductivity, *Phys. Rev.*, **132**(6), pp. 2461–2471, doi:10.1103/PhysRev.132.2461.
9. Klemens, P. G. (1966). Anharmonic decay of optical phonons, *Phys. Rev.*, **148**(2), pp. 845–848, doi:10.1103/PhysRev.148.845.
10. Debernardi, A., Baroni, S., and Molinari, E. (1995). Anharmonic phonon lifetimes in semiconductors from density-functional perturbation theory, *Phys. Rev. Lett.*, **75**(9), pp. 1819–1822, doi:10.1103/PhysRevLett.75.1819.
11. Rowlette, J., and Goodson, K. (2008). Fully coupled nonequilibrium electron–phonon transport in nanometer-scale silicon fets, *IEEE Trans. Electron Devices*, **55**(1), pp. 220–232, doi:10.1109/TED.2007.911043.

12. Jacoboni, C., and Reggiani, L. (1983). The Monte Carlo method for the solution of charge transport in semiconductors with applications to covalent materials, *Rev. Mod. Phys.*, **55**(3), pp. 645–705.

13. Reggiani, L. (ed.) (1984). *Hot-Electron Transport in Semiconductors*, Springer-Verlag, Berlin.

14. Hess, K. (2000). *Advanced Theory of Semiconductor Devices*, IEEE Press, New York, NY.

15. Hjelm, M., and Nilsson, H. E. (2002). Methods for anisotropic selection of final states in the full band ensemble Monte Carlo simulation framework, *Simulat. Pract. Theory*, **9**, pp. 321–332.

16. Klemens, P. G. (1967). Decay of high-frequency longitudinal phonons, *J. Appl. Phys.*, **38**(12), pp. 4573–4576, doi:10.1063/1.1709187, http://link.aip.org/link/?JAP/38/4573/1.

17. Seitz, F., and Turnbull, D. (eds.) (1970). *Solid State Physics*, Vol. 7, Academic Press, New York, NY.

18. Balandin, A., and Wang, K. L. (1998). Significant decrease of the lattice thermal conductivity due to phonon confinement in a free-standing semiconductor quantum well, *Phys. Rev. B*, **58**(3), pp. 1544–1549, doi:10.1103/PhysRevB.58.1544.

19. Weber, W. (1977). Adiabatic bond charge model for the phonons in diamond, Si, Ge, and α-Sn, *Phys. Rev. B*, **15**, pp. 4789–4803.

20. Nielsen, O. H., and Weber, W. (1979). Lattice dynamics of group IV semiconductors using an adiabatic bond charge model, *Comp. Phys. Commun.*, **18**, pp. 101–107.

21. Srivastava, G. P. (1990). *The Physics of Phonons*, Taylor and Francis, New York, NY.

22. Ziman, J. M. (1960). *Electrons and Phonons: The Theory of Transport Phenomena in Solids*, Oxford University Press, Oxford, UK.

Chapter 6

# Phonon Monte Carlo: Generating Random Variates for Thermal Transport Simulation

**L. N. Maurer, S. Mei, and I. Knezevic**
*University of Wisconsin-Madison, Madison, WI 53706, USA*
lnmaurer@wisc.edu, smei4@wisc.edu, iknezevic@wisc.edu

## 6.1 Introduction

Thermal transport in semiconductors is governed by phonons, the quanta of lattice waves [1, 2]. At temperatures above several degrees Kelvin, phonons experience considerable multiphonon interactions, in addition to scattering with boundary roughness, interfaces, or atoms of different masses that stem from doping, alloying, or natural isotope variation. Therefore, phonon transport in nanostructures is typically in the quasi-ballistic or diffusive regimes and is described well by the phonon Boltzmann transport equation (PBTE). The PBTE can be solved via deterministic [3, 4] or stochastic techniques [5–7].

Phonon Monte Carlo (PMC) is an efficient stochastic technique for the solution to the PBTE [5–15], which can incorporate real-

*Nanophononics: Thermal Generation, Transport, and Conversion at the Nanoscale*
Edited by Zlatan Aksamija

ISBN 978-981-4774-41-3 (Hardcover), 978-1-315-10822-3 (eBook)
www.panstanford.com

space roughness and simulate nanostructures of experimentally relevant sizes. In PMC, a large ensemble of numerical phonons (typically on the order of $10^5$ – $10^6$ ) is tracked over time as they fly freely and undergo scattering according to relevant scattering rates [16]. Modern transport simulations based on the PBTE involve phonon-phonon scattering rates obtained from first principles [3, 17] and incorporate full phonon dispersions [12]; the latter are very important in anisotropic systems, such as superlattices, nanowires, and nanoribbons [12, 18].

A number of random variables with generally complicated distribution functions underscore the behavior of the phonon ensemble. Examples of random variables are the phonon energy or momentum for a bulk phonon system in equilibrium and the outgoing momentum direction after a scattering event. The PMC simulation hinges on the generation and use of random variates—specific values of the random variables that correspond to physical observables—in a way that accurately and efficiently captures the appropriate distribution functions. Accurate and efficient generation of random variates that numerically represent nonuniform distribution functions is not a simple matter, yet most articles on PMC do not show much detail on this aspect.

In this chapter, we discuss numerical generation of random variates relevant for PMC, assuming that the uniformly distributed variates on the [0, 1] interval are accessible. We discuss the relative merits of different approaches (direct inversion versus the rejection technique) from both theoretical and practical standpoints, which are sometimes at odds, and show several specific examples of nonuniform distributions relevant for phonon transport. We also identify common themes in phonon generation and scattering that are useful for reusing code in a simulation. We trust these examples will inform the reader about both the mechanics of random-variate generation and how to choose a good approach for whatever problem is at hand.

We review the two main methods for generating random variates in Section 6.2 and the PMC method in Section 6.3. Several applications are presented next: generating the attributes for phonons in equilibrium with full (Section 6.4.1) and isotropic

(Section 6.4.2) dispersions, randomizing outgoing momentum upon diffuse boundary scattering (Section 6.5), implementing contacts (Section 6.6), and conserving energy in the simulations (Section 6.7).

## 6.2 Generating Random Variates

Here, we present a quick overview of the methods used to generate random variates with a given probability distribution function (PDF). We assume that the computer used can generate random variates that are uniformly distributed over the interval [0, 1].

There are two techniques for generating nonuniform random variates from uniform random variates, the inversion method and the rejection method, which we explain later. (For more details on random-variate generation, see a book such as Ref. [19].)

Generally speaking, the inversion method requires more analytical manipulation of the PDF than the rejection method. When the analytical manipulations are possible, the inversion method is usually the simpler of the two. While the inversion method can be performed numerically, the rejection technique is generally simpler to implement in cases when analytical inversion is not possible.

### 6.2.1 The Inversion Method

Consider a PDF $p(x)$. The first step in the inversion method is to integrate the PDF into the cumulative distribution function (CDF),

$$F(x) = \int_{-\infty}^{x} dx' f(x'), \tag{6.1}$$

where $F(x)$ is the probability that a random variate will have a value less than or equal to $x$.

Next, we generate a random variate $r$, which is uniformly distributed in [0, 1]. We solve $r = F(x)$ for $x$, that is, we invert the CDF to get the quantile function $Q(r) = F^{-1}(r)$. Finally, we solve $x = Q(r)$ for $x$. The resulting $x$ is a random variate that follows our original PDF [19]. The technique can be generalized to PDFs with multiple variables, but we will only consider PDFs that effectively only depend on a single variable.

For example, say we want to generate a random variate from the distribution given by Lambert's cosine law[a] in 3D, $p(\theta, \phi) = c \cos\theta$ for spherical coordinates $\theta \in [0, \pi/2]$, $\phi \in [0, 2\pi)$ and a normalizing constant $c$ [20]. Lambert's cosine law will prove important later (Sections 6.5 and 6.6.2). First, we must properly normalize $p(\theta, \phi)$:

$$1 = \int_0^{\pi/2} \int_0^{2\pi} p(\theta', \phi') \sin\theta' d\theta' d\phi', \tag{6.2}$$

which yields $c = \pi^{-1}$. The CDF for $\theta$ is then

$$F(\theta) = \int_0^{\theta} \int_0^{2\pi} p(\theta', \phi') \sin\theta' \, d\theta' \, d\phi' \tag{6.3}$$

$$= \sin^2\theta.$$

Note that our CDF only depends on $\theta$. We can do this since $p(\theta, \phi)$ does not depend on $\phi$, but we keep the $\phi$ dependence to clarify that we still need to integrate over $\phi$. Finally, invert the CDF for a random variate $r_\theta$ that is uniformly distributed in [0, 1]:

$$F(\theta) = r_\theta.$$

$$\theta = \arcsin\left(\sqrt{r_\theta}\right). \tag{6.4}$$

Using the same method, we can find the unsurprising result that $\phi = 2\pi r_\phi$, where $r_\phi$ is uniformly distributed in [0, 1].

In this example, both integration and inversion can be done analytically, which is the exception rather than the rule. It is possible to do both integration and inversion numerically, but this reduces the accuracy and computational efficiency of the method. We will see cases where both integration and inversion are done numerically (Section 6.4.1) and where integration can be done analytically but inversion is done numerically (Section 6.5).

### 6.2.2 The Rejection Method

In contrast to the inversion method, the rejection method does not require any integration or inversion steps. The rejection method requires that when we calculate the PDF $p(x)$ for any $x$, there exists a bounding function $g(x)$ such that $\forall x, g(x) \geq p(x)$, and that we can generate random variates from a PDF that is proportional to $g(x)$. $p(x)$ and $g(x)$ do not have to be normalized; they must simply be proportional to PDFs. The largest drawback of the rejection method

[a]This is also known as Knudsen's cosine law in the context of gas molecules scattering from surfaces.

is that, unlike the inversion method, the rejection method generally requires the computer to generate several random variates that are uniformly distributed in [0, 1]. Choosing a $g(x)$ that closely resembles $p(x)$ will reduce the number of random variates that the computer will have to generate.

The rejection method to generate a random variate from $p(x)$ follows.

(1) Generate a random variate $x'$ from the PDF that is proportional to $g(x)$.

(2) Generate a random variate $y$ that is uniformly distributed in $[0, g(x')]$.

(3) If $y < p(x')$, then $x'$ is the random variate generated from $p(x)$. Otherwise, return to step 1.

The third step ensures that the probability of choosing $x'$ is proportional to $p(x')$, which is all that we require of a method to generate random variates from a distribution.

This procedure is easy to visualize if $p(x)$ is nonzero only on a finite interval $[a, b]$, and $p(x) \leq c$, where $a$, $b$, $c$ are constants (Fig. 6.1). In this case, we can choose $g(x) = c$. Then the rejection method is equivalent to throwing a dart randomly and uniformly at the box defined by $x' \in [a, b]$, $y \in [0, c]$. If $y \leq p(x')$, that is, the dart falls below the curve $p(x)$, then we take $x'$ as our random variate and repeat the procedure otherwise. The larger the area between the line $y = c$ and the curve $p(x)$, the more dart throws will be required for a dart to land below the curve $p(x)$, which reduces the efficiency of the rejection method. For this reason, even if $p(x) \leq c$, it may be wise to use a $g(x)$ other than $g(x) = c$. We do this in Section 6.5 when we consider Soffer's model for momentum-dependent boundary scattering [21].

We note that the rejection technique can work even if $p(x)$ diverges, which is common in physics problems (e.g., Van Hove singularities). Take the case of a probability distribution that might arise from the Bose–Einstein distribution in 2D and in polar coordinates:

$$p(r, \theta) = \frac{1}{e^r - 1}. \tag{6.5}$$

Note that $p(r, \theta)$ is not normalized and diverges at $r = 0$. Also note that $p(r, \theta) \leq r^{-1}$, so we choose $g(r, \theta) = r^{-1}$. Finally, suppose that we are considering the 2D domain $\theta \in [0, 2\pi]$, $r \in [0, R]$, where $R$ is a constant. On that domain, $g(r, \theta)$ and $p(r, \theta)$ are normalizable because the singularity at $r = 0$ is integrable: $\int_0^{2\pi}\int_0^R g(r, \theta)\, rdrd\theta = 2\pi R$, which is finite. Because the integrand $g(r, \theta)\, r$ is constant, the random variate $r'$ is equally likely to take any value in $[0, R]$. So, it is quite simple to generate the random variates $r'$, and $\theta'$ : $r'$ is uniformly distributed in $[0, R]$ and $\theta'$ is uniformly distributed in $[0, 2\pi]$.

Putting everything together, the rejection technique in this example works as follows:

(1) Generate random variates $r'$ and $\theta'$ that are uniformly distributed in $[0, R]$ and $[0, 2\pi]$, respectively.

(2) Generate a random variate $y$ that is uniformly distributed in $[0, (r')^{-1}]$.

(3) If $y < p(r', \theta') = (e^{r'} - 1)^{-1}$, then use $r'$ and $\theta'$ as your random variates. Otherwise, return to step 1.

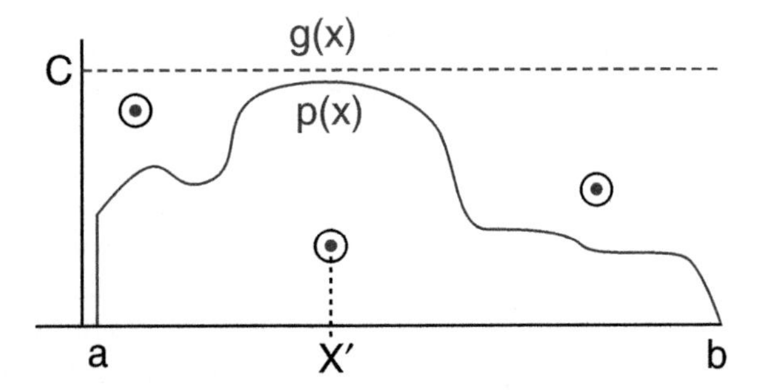


**Figure 6.1** Illustration of the rejection method for a distribution $p(x)$ using a constant $g(x)$. For a constant $g(x)$, the rejection technique is the same as randomly throwing darts uniformly in the range $x \in [a, b]$, $y \in [0, c]$. If the dart lands above the curve $p(x)$, then another dart is thrown. When a dart lands below the curve $p(x)$, then its $x$ value, $x'$, is the random variate that is accepted. The figure depicts possible dart throws (marked with bulls-eyes). The throws above $p(x)$ are rejected, and $x'$ is taken from the throw that lands below $p(x)$.

## 6.3 Overview of Phonon Monte Carlo

Phonons are the main carriers of heat in semiconductor materials [2]. Phonons can be treated as semiclassical particles on the spatial scales longer than the phonon coherence length and timescales longer than the phonon relaxation time. Phonon transport under these conditions is captured via the PBTE [1]:

$$\frac{\partial n_{\mathrm{b}}(\mathbf{r},\mathbf{q},t)}{\partial t}+v_{\mathrm{b},\mathbf{q}}\cdot\nabla_{\mathbf{r}}n_{\mathrm{b}}(\mathbf{r},\mathbf{q},t)=\left.\frac{\partial n_{\mathrm{b}}(\mathbf{r},\mathbf{q},t)}{\partial t}\right|_{\mathrm{scat}}. \tag{6.6}$$

$n_{\mathrm{b}}(\mathbf{r},\mathbf{q},t)$ is the time-dependent distribution of phonons with respect to position $\mathbf{r}$ and the phonon wave vector $\mathbf{q}$ for phonon branch b. $\mathbf{v}_{\mathrm{b},\mathbf{q}} = \nabla\mathbf{q}\ \omega_{\mathrm{b},\mathbf{q}}$ is the phonon group velocity, where $\omega_{\mathrm{b},\mathbf{q}}$ is the phonon angular frequency in branch b at wave vector $\mathbf{q}$. In equilibrium, the average occupancy of a phonon state with energy $\hbar\omega$ at absolute temperature $T$ is given by the Bose–Einstein distribution function

$$\langle n_{\mathrm{BE}}(\omega,T)\rangle=\frac{1}{e^{\frac{\hbar\omega}{k_{\mathrm{B}}T}}-1}, \tag{6.7}$$

where $k_B$ is the Boltzmann constant. When addressing out-of-equilibrium phonon transport, we assume local equilibrium, employ the concept of a local and instantaneous temperature, $T(\mathbf{r},t)$, and then calculate the expectation number of phonons accordingly.

To obtain the phonon group velocity $\mathbf{v}_{\mathrm{b},\mathbf{q}}$, we require the phonon dispersion relations—the relationships between the phonon angular frequency $\omega_{\mathrm{b},\mathbf{q}}$ and the phonon wave vector $\mathbf{q}$. Figure 6.2 shows the full phonon dispersions along the high-symmetry directions in 3D silicon and 2D graphene. In acoustic branches, the phonon dispersion is relatively isotropic. Therefore, an analytical isotropic phonon dispersion ($\omega_{\mathrm{b},\mathbf{q}}$ depending only on the magnitude of $\mathbf{q}$) is often adopted as it enables a direct and simple evaluation of the phonon group velocity from the random phonon wave vector obtained after scattering (more on this in Section 6.2), which speeds up the simulation. A quadratic isotropic dispersion[b] has been shown to be very accurate in simulating the thermal properties of silicon [6, 8, 10].

[b] $\omega_{\mathrm{b}}(\mathbf{q}) = v_{s,\mathrm{b}}\,q + c_{\mathrm{b}}\,q^2$, where $q = |\mathbf{q}| \in [0, q_{\max}]$ is the allowed wave number, $v_{s,\mathrm{b}}$ is the mode-dependent sound velocity, and $c_{\mathrm{b}}$ is the quadratic coefficient fitted from the full dispersion. The maximum allowed wave number $q_{\max}$ is calculated from $\frac{2p}{a_0}$, where $a_0$ is the lattice constant of the material.

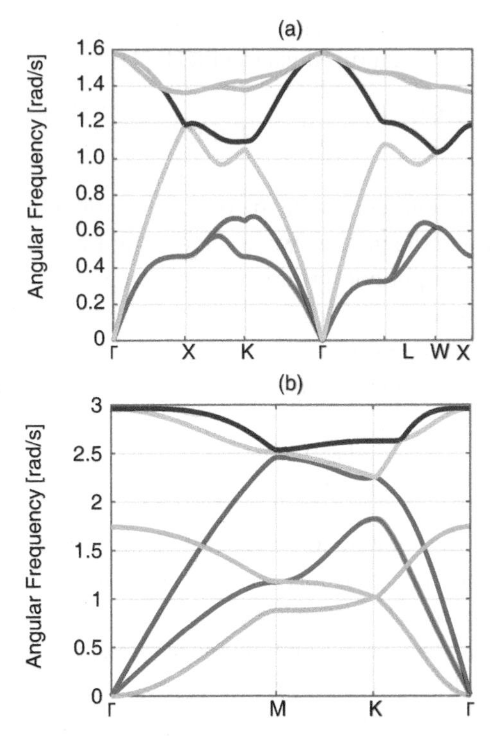


**Figure 6.2** Phonon dispersions for (a) 3D silicon and (b) 2D graphene along high-symmetry directions.

However, when the simulated structure is very small, such as in the case of thin nanowires or graphene nanoribbons (GNRs), anisotropy is prominent and adopting the full dispersion is necessary at the cost of slowing down the simulation. In the following sections, the isotropic approximation and the full dispersion relations in the context of PMC will be demonstrated on the examples of 3D silicon and 2D graphene, respectively.

PMC is a widely used stochastic technique for solving the PBTE and obtaining thermal transport properties of semiconductor materials [6, 8]. PMC tracks phonon transport in real space, thus allowing easy implementation of nontrivial geometries such as rough boundaries and real-space edge structures [10, 12, 14]. Figure 6.3 shows the flowchart of a typical PMC simulation. The simulation domain is a wire with a square cross section (3D) or a rectangle (2D) divided into $N_c$ cells of equal length along the heat transport direction, as shown in Fig. 6.4. The two end cells are connected to heat reservoirs fixed at slightly different temperatures ($T_h$ and

$T_c$). For structures much longer than the phonon mean free path (i.e., in the diffusive transport limit) and in a steady state, a linear temperature profile between $T_h$ and $T_c$ will naturally develop inside the simulation domain (Fig. 6.4). Therefore, we can achieve a steady state in the simulation faster if we initialize the cells according to a linear temperature profile, with the $i$th cell temperature set to $T_i = T_h - \frac{i-1}{N_c-1}(T_h - T_c)$.

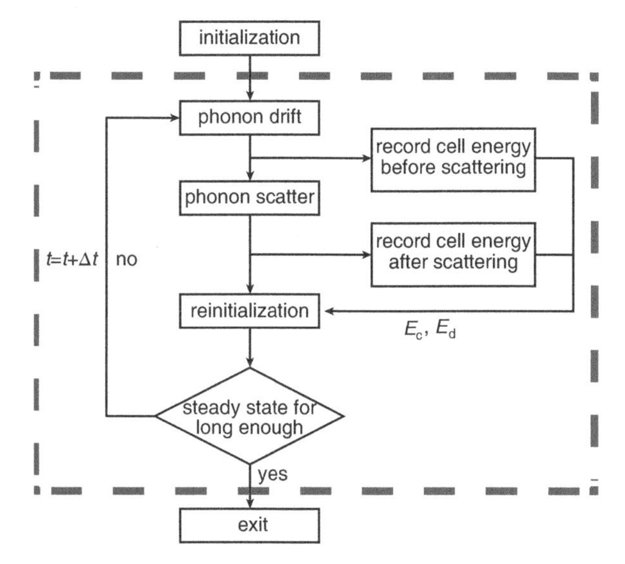


**Figure 6.3** Flowchart of a PMC simulation. The dashed box encloses the transport kernel.

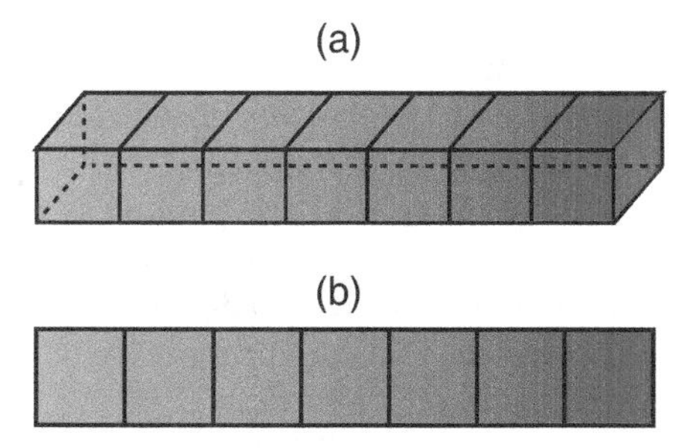


**Figure 6.4** A typical simulation domain for (a) a 3D and (b) a 2D PMC simulation. The color represents a typical steady-state temperature profile in each structure when connected to cold (blue, left end) and hot (red, right end) reservoirs.

The expected total energy associated with the $i$th cell is then

$$\mathcal{E}_i = \Omega_i \sum_{\mathrm{b}} \int D_{\mathrm{b}}(\omega) \langle n_{\mathrm{BE}}(\omega, T_i) \rangle \hbar\omega \, d\omega, \tag{6.8}$$

where $\Omega_i$ is the volume (3D) or area (2D) of the $i$th cell in real space, $D_{\mathrm{b}}(\omega)$ is the phonon density of states (PDOS) in the material associated with branch b and energy level $\hbar\omega$, and the sum is over all three acoustic branches. The expectation number of phonons in the $i$th cell is then

$$\mathcal{N}_{i,\mathrm{exp}} = \Omega_i \sum_{\mathrm{b}} \int D_{\mathrm{b}}(\omega) \langle n_{\mathrm{BE}}(\omega, T_i) \rangle \, d\omega. \tag{6.9}$$

In simulations where the sample size is large (on the order of microns) or the temperature is not very low (a few hundred Kelvin), the expectation number of phonons $\mathcal{N}_{i,\mathrm{exp}}$ can be very high ($10^7$ – $10^9$) and it is computationally expensive to keep track of this many particles in the simulation. Instead, a weighting factor $W$ is often introduced [6, 8] to reduce the number of simulation particles to a tolerable range (typically $10^5$ – $10^6$).

In 2D materials, like graphene, the number of flexural out-of-plane (ZA) phonons is overwhelmingly larger than that of transverse acoustic (TA) or longitudinal acoustic (LA) phonons, owing to the shape of the dispersion curves; therefore, a branch-dependent weighting factor $W_{\mathrm{b}}$ should be used. With the weighting factor taken into consideration, the total energy of the simulation particles in a cell becomes

$$E_i = \Omega_i \sum_{\mathrm{b}} \int D_{\mathrm{b}}(\omega) \frac{\langle n_{\mathrm{BE}}(\omega, T_i) \rangle}{W_{\mathrm{b}}} \hbar\omega \, d\omega. \tag{6.10}$$

During the initialization, we keep generating phonons following a desired distribution according to each cell's temperature and add them to random positions inside the cell until the cell has the desired energy, as expressed by Eq. 6.10. After initialization, we enter the transport kernel (enclosed in the dashed box in Fig. 6.3), where time is discretized in steps of $\Delta t$. Each phonon is allowed to drift according to the group velocity obtained through the dispersion relation. A random number is drawn to decide whether the phonon will be scattered during its drift; the probability of being scattered is captured through a phonon relaxation time. We record the heat flux along the wire (3D)/ribbon (2D) to monitor whether a steady state has been reached (i.e., whether the flux has become constant) and use ensemble averages to compute the thermal properties of the material. Special measures (e.g., reinitialization) are needed to make sure the energy in each cell is properly conserved without violating the distribution (Section 6.7). More details about the full dispersion PMC simulation can be found in [12].

## 6.4 Generating Phonon Attributes in PMC

### 6.4.1 Thermal Phonons with Full Dispersion in 2D

With the basic knowledge of the methods to generate random variates following certain distributions, this section gives examples of using these methods to randomly draw a phonon from the equilibrium distribution, following the full dispersion relation in 2D graphene.

Here, we will present both the inversion technique with numerical integration and inversion and the rejection technique; the former is commonly used for thermal phonons with isotropic dispersion in 3D [6, 7, 10], as is the rejection technique [5, 14]. The inversion technique is used to choose the phonon frequency and branch, and the rejection technique is used to choose a wave vector that matches the chosen frequency and branch. We note that we could use the rejection technique to choose the branch and wave vector directly (the method would be similar to the example at the end of Section 6.2.2), but we choose a hybrid approach to allow for code reuse: for internal scattering, we already need the code that can generate phonons of a specific frequency. So, we use the inversion technique to choose the phonon frequency and then use the pre-existing code to draw a corresponding wave vector.

As introduced in Section 6.3, we have assigned a temperature to the cell that we are generating a phonon in. The first step in generating the phonon is to find an angular frequency that follows the Bose–Einstein distribution according to the temperature. To do that, we use the CDF of $\omega$ at the given temperature $T$ with the help of the density of states (DOS):

$$F(\omega,T)=\frac{\sum_b \int_0^{\omega} d\omega \langle n_{\mathrm{BE}}(\omega,T)\rangle D_{\mathrm{b}}(\omega)/W_{\mathrm{b}}}{\sum_b \int_0^{\omega_{\max}} d\omega \langle n_{\mathrm{BE}}(\omega,T)\rangle D_{\mathrm{b}}(\omega)/W_{\mathrm{b}}}. \tag{6.11}$$

With full dispersion, we do not have an analytical expression for $D_b(\omega)$. Therefore, we divide the range of $[0, \omega_{max}]$ into $N_{int}$ equal-energy bins where $\Delta\omega = \left\lceil \frac{\omega_{max}}{N_{int}} \right\rceil$ is the interval length and $\omega_{c,i} = \frac{2i-1}{2}\Delta\omega$ is the central frequency of the $i$th interval. We can obtain $D_b(\omega_{c,i})$ for $i = 1, 2, \ldots, N_{int}$ and evaluate the discrete CDF as

$$F_i(T) = \frac{\sum_b \sum_{j=1}^{i} \langle n_{BE}(\omega_{c,j}, T) \rangle D_b(\omega_{c,j}) / W_b}{\sum_b \sum_{j=1}^{N_{int}} \langle n_{BE}(\omega_{c,j}, T) \rangle D_b(\omega_{c,j}) / W_b}. \quad (6.12)$$

To complete the table, set $F_0(T) = 0$, meaning that all phonons must have positive energy. The numerically evaluated CDF for phonons at 300 K is shown in Fig. 6.5.

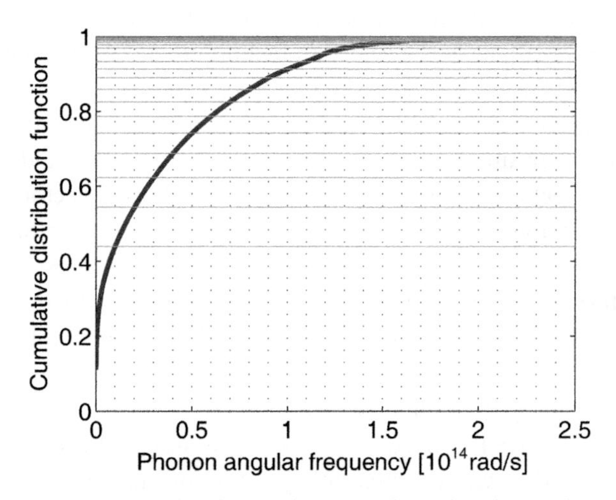


**Figure 6.5** Cumulative distribution function of angular phonon frequency at 300 K. The frequency range $\omega \in [0, 2.5 \times 10^{14}]$ rad/s was divided into $N_{int} = 2500$ equal intervals in the numerical calculation. Reproduced from from Ref. [12], with the permission of AIP Publishing.

The next step is to numerically invert the table: we draw a random number $r_1$ and look for the interval $i$ satisfying $F_{i-1} < r_1 < F_i$.

We decide the frequency of this phonon falls in the $i$th interval and the actual frequency is determined with another random number $r_2$,

$$\omega = \omega_{c,i} + (2r_2 - 1)\,\frac{\Delta\omega}{2}.$$

When the $\omega$ is chosen, phonon branch b can be chosen in a similar fashion. Use indexes 1, 2, and 3 to represent the TA, LA, and ZA branches, respectively. Since there are only three branches, we can enumerate the CDFs as

$$F_1(\omega) = \frac{D_1(\omega)/W_1}{\sum_{b'} D_{b'}(\omega)/W_{b'}}, \tag{6.13a}$$

$$F_2(\omega) = \frac{D_1(\omega)/W_1 + D_2(\omega)/W_2}{\sum_{b'} D_{b'}(\omega)/W_{b'}}, \tag{6.13b}$$

$$F_3(\omega) = 1, \tag{6.13c}$$

and a third random number $r_3$ is used to choose the phonon branch b.

The next step is generating the phonon wave vector, $\mathbf{q}$, for the $\omega$ and b we already found. The distribution is 2D and complicated, so it is hard to calculate a CDF. As a result, we employ the rejection technique. Figure 6.6 shows the isoenergy curves for the TA branch in the first Brillouin zone (1BZ), adjacent curves differing by $2 \times 10^{13}$ rad/s. Because the isoenergy curves are close to circles, it is convenient to use the polar coordinate. Further, because of the symmetry, we can generate $\mathbf{q}$ in the shaded area and simply map it to the whole 1BZ. (More details can be found in Ref. [12].) We use the rejection technique to choose an angle $\theta \in \left[0, \frac{\pi}{6}\right]$ and use a lookup table to get the corresponding $|\mathbf{q}|$. The probability of a phonon with frequency $\omega$ in branch b having angle $\theta$ is represented by

$$p(\omega,\theta) \propto \frac{arc(\theta-\delta\theta,\theta+\delta\theta)}{\left|v_g(\omega,\theta)\right|}, \tag{6.14}$$

where $arc(\theta - \delta\theta, \theta + \delta\theta)$ is the arc length on the isoenergy curve between $(\theta - \delta\theta, \theta + \delta\theta)$ and $|\mathbf{v}_g(\omega, \theta)|$ is the magnitude of group velocity. When $\delta\theta$ is small, it is acceptable to assume the group velocity is constant along the arc. Since we do not have an analytical expression for the probability, we make a rejection table of $N_a$

equally separated points between $\left[0, \frac{\pi}{6}\right]$, where $\Delta\theta = \frac{\theta_{\max}}{N_a}$ is the spacing and $\theta_{c,i} = (2i - 1)\frac{\Delta\theta}{2}$ is the central frequency in the $i$th interval. Note that energy is also discrete; we evaluate Eq. 6.14 only at $(\omega_{c,i}, \theta_{c,j})$ and obtain an $N_{\text{int}} \times N_a$ interpolation table. For any $0 < \omega < \omega_{\max}$ and $0 < \theta < \theta_{\max}$ we can get the probability of having a phonon from interpolation. $N_a = 100$ is enough for accurate interpolation. Upon getting the energy $\omega$, we get the $1 \times N_a$ angle-distribution table $f_\omega(\theta_{c,i})$ and record $f_{\omega,\max} = \max_{i=0}^{N_a}(f_\omega(\theta_{c,i}))$. A pair of uniformly distributed random numbers $(x, y)$ is drawn. Therefore, our angle in consideration is $\alpha = \frac{\pi}{6} \cdot x$. Then interpolate the distribution table to obtain $f_\omega(\alpha)$. If $f_{\omega,\max} \cdot y \le f_\omega(\alpha)$, the angle $\alpha$ is accepted, and we can proceed to look up $|\mathbf{q}|(\omega, \alpha)$ by interpolation. The final chosen wave vector is then $(|\mathbf{q}|(\omega, \alpha)\cos\alpha, |\mathbf{q}|(\omega, \alpha)\sin\alpha)$. If $f_{\omega,\max} \cdot y > f_\omega(\alpha)$, the angle is rejected. We keep generating $(x, y)$ pairs until an angle is accepted, which typically occurs within two iterations with the simple $f_{\omega,\max}$ bound.

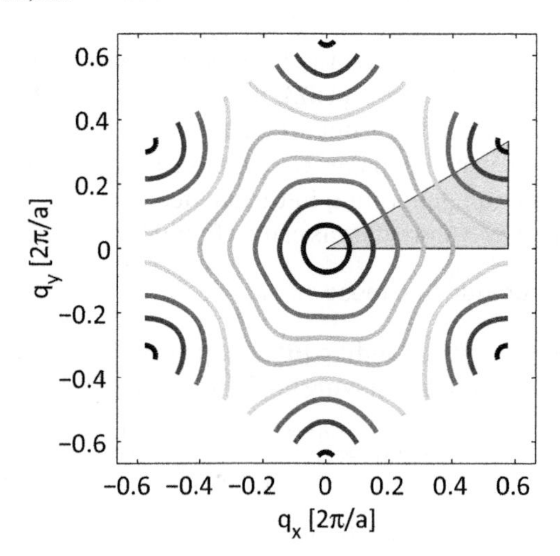


**Figure 6.6** Isoenergy curves in the first Brillouin zone for TA branch phonons. Adjacent curves are offset by $2 \times 10^{13}$ rad/s, and the shaded region in the black triangle is the irreducible wedge in the first Brillouin zone. Reproduced from from Ref. [12], with the permission of AIP Publishing.

### 6.4.2 Thermal Phonons with an Isotropic Dispersion in 3D

As we mentioned in the previous section, it is common to draw thermal phonons with isotropic dispersion relations in 3D using the inversion technique with numerical integration and inversion [6, 7, 10], although the rejection method is also sometimes used [5, 14]. We believe that the rejection method is better suited to the task than the inversion method because the numerical integration and inversion can lead to a loss of accuracy or decreased computational performance [6]. Additionally, the rejection method does not require the DOS, which can be difficult to calculate. We will consider two branches, a TA branch and an LA branch, but the method can easily be generalized to more branches. We will also assume $q_{\max} \geq |\mathbf{q}|$ is an upper bound on the wave vector magnitude. The rejection method works as follows.

At temperature $T$, the number of phonons from branch b in each infinitesimal unit of reciprocal space is

$$n_{\mathrm{b}}(\mathbf{q},T)=\frac{d^3\mathbf{q}}{(2\pi)^3}\langle n_{\mathrm{BE}}(\omega_{\mathrm{b},\mathbf{q}},T)\rangle. \tag{6.15}$$

For an isotropic dispersion relation, $n_{\mathrm{b}}$ can be reduced to a function of $q$ alone by integrating over the polar and azimuthal angles $\theta$ and $\phi$.

$$\begin{aligned} n_{\mathrm{iso,b}}(q,T) &= \int_0^{2\pi}\int_0^{\pi}\frac{q^2\sin\theta dqd\theta d\phi}{(2\pi)^3}\langle n_{\mathrm{BE}}(\omega_{\mathrm{b},\mathbf{q}},T)\rangle \\ &= \frac{q^2}{2\pi}\langle n_{\mathrm{BE}}(\omega_{\mathrm{b},q},T)\rangle dq. \end{aligned} \tag{6.16}$$

Although $\langle n_{BE}\rangle$ diverges at $q = 0$, $\lim_{q\to 0} n_{\mathrm{b}}(\mathbf{q}) = 0$ because of the $q^2$ term. Thus, the maximum value of $n_{\mathrm{iso,b}}(q, T)$ is finite, and we can use the bounding function $g(q) = c$, as described in Section 6.2.2, where $c$ is any number greater than the maximum value of $n_{\mathrm{iso,TA}}(q, T) + n_{\mathrm{iso,LA}}(q, T)$. The maximum value of $n_{\mathrm{b}}(q, T)$ is temperature dependent, but instead of finding a new $c$ whenever the temperature changes, simply find a $c$ that works for a temperature higher than any conceivable temperature in your simulation. Once $c$ is found, the rejection method follows the familiar pattern:

(1) Generate a random variate $q'$ that is uniformly distributed in $[0, q_{\max}]$.

(2) Generate a random variate $y$ that is uniformly distributed in $[0, c]$.

(3) If $y < n_{\text{iso,TA}}(q, T)$, then generate a TA phonon with wave number $q'$. If $n_{\text{iso,TA}}(q', T) < y < n_{\text{iso,TA}}(q', T) + n_{\text{iso,LA}}(q', T)$, then generate an LA phonon with wave number $q'$. Otherwise, return to step 1.

Choosing between different branches in the last step is similar to the procedure used for choosing a branch in Eq. 6.13.

Once we know the wave number $q$ of the new phonon, we need to choose a direction. In equilibrium, all directions are equally likely [$p(\theta, \phi) = d$, where $d$ is a constant] and we can use the inversion technique to choose a direction. First, we find $d$.

$$
\begin{aligned}
1 &= \int_0^{\pi}\int_0^{2\pi} p(\theta',\phi')\sin\theta' d\theta' d\phi' \\
d &= (4\pi)^{-1}.
\end{aligned}
\tag{6.17}
$$

The CDF is[c]

$$
\begin{aligned}
P(\theta) &= \int_0^{\theta}\int_0^{2\pi} p(\theta',\phi')\sin\theta' d\theta' d\phi' \\
&= \frac{1-\cos\theta}{2}.
\end{aligned}
\tag{6.18}
$$

Inverting for a random variate $r_\theta$ that is uniformly distributed in [0, 1]

$$
\theta = \arccos(1 - 2r_\theta). \tag{6.19}
$$

$\phi$ is uniformly distributed in $[0, 2\pi]$.

All together

$$
\mathbf{q} = q' \begin{pmatrix} \sin(\theta)\cos(\phi) \\ \sin(\theta)\sin(\phi) \\ \cos(\theta) \end{pmatrix}. \tag{6.20}
$$

## 6.5 Diffuse Boundary Scattering

Nanostructure surfaces can have a large impact on thermal conductivity through increased boundary-surface scattering [22],

[c]In the CDF, $(1 - \cos\theta)/2$ can be rewritten as $\sin^2(\theta/2)$, but making that change offers no advantage in implementing the code, although it does make the equation look more like the results from Lambert's cosine law (Eq. 6.4).

so implementing real-space phonon-surface scattering is important for PMC simulations. A common class of phonon-surface scattering models relies on diffuse scattering and a specularity parameter: a phonon that strikes a surface either reflects specularly or is scattered into a new phonon of the same frequency and branch, but with a random wave vector. The probability of specular scattering is controlled by a specularity parameter, which is either constant [23] or momentum dependent [1, 21, 24]. Specularity parameter models fail when the surfaces are very rough but appear to work well for small roughnesses [14]. Additionally, real phonons of one branch can be scattered into phonons of a different branch at a surface [25]. This process, referred to as mode conversion, is often ignored but can be included in PMC simulations [26].

All phonon-surface scattering models must respect detailed balance in equilibrium: the number of phonons scattered into any solid angle must be equal to the number of phonons incident on the surface from the same solid angle. Consider a flat surface with a surface normal vector $\hat{n}$. In this section, $\theta$ will refer to the angle between a phonon's wave vector $\mathbf{q}$ and the surface's unit normal vector. (We will assume that the dispersion relation is isotropic so that $\mathbf{q}$ is in the same direction as the velocity.) Phonons with a larger $\mathbf{q} \cdot \hat{n} = |\mathbf{q}| \cos \theta$ will strike the surface more frequently, and in equilibrium this leads to Lambert's cosine law: the number of phonons incident on a surface at an angle $\theta$ must also be proportional to $\cos \theta$ (Fig. 6.7). Because of detailed balance, the number of phonons being scattered into an angle $\theta$ is proportional to $\cos \theta$.

For specular scattering, detailed balance in equilibrium holds automatically. For totally diffuse scattering, the inversion method can be used to generate random wave vectors that satisfy Lambert's cosine law. Lambert's cosine law in 3D was an example in Section 6.2.1, with the result given in Eq. 6.4. In 2D, the procedure is very similar except that we work in polar rather than spherical coordinates. We use the distribution $p(\theta) = c \cos\theta$, where $c$ is a normalizing constant. The normalization condition is now

$$1 = \int_0^{\pi/2} p(\theta')\, d\theta', \tag{6.21}$$

which yields $c = 1$. The CDF is

$$\begin{aligned} F(\theta) &= \int_0^{\theta} p(\theta')d\theta' \\ &= \sin\theta. \end{aligned} \tag{6.22}$$

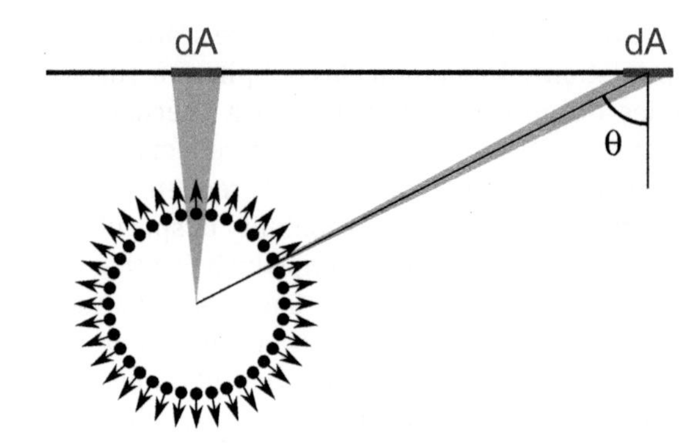


**Figure 6.7** Illustration of Lambert's cosine law. An equilibrium ensemble of phonons is depicted as a group of particles with uniform angular distribution (lower left). The number of phonons striking a small patch of surface *dA* at an angle $\vartheta$ (measured from the surface normal) is proportional to the angle subtended by the shaded wedges. This angle is proportional to cos $\vartheta$ (if *dA* is small), which is the basis for Lambert's cosine law.

Finally, we invert the CDF for a random variate $r_\theta$ that is uniformly distributed in [0, 1]:

$$F(\theta) = r_\theta,$$

$$\theta = \arcsin \theta, \tag{6.23}$$

which resembles the 3D result.

For the constant specularity parameter model in both 2D and 3D, one simply chooses a probability $p$ of specular scattering before running the simulation. Then, each time a phonon strikes the surface, the phonon is specularly scattered with probability $p$ and is otherwise diffusely scattered into a randomly chosen direction using Eq. 6.4 or Eq. 6.23 for 3D or 2D, respectively.

The situation is somewhat more complicated for momentum-dependent specularity parameter models. Take Soffer's model in 3D [21]. The probability that a phonon scatters specularly is

$$p(\theta, \phi) \propto e^{-(2\sigma|\mathbf{q}| \cos \theta)^2}, \tag{6.24}$$

where $\sigma$ is the root mean square (RMS) roughness of the surface. Because the probability of diffuse scattering now depends on the angle of incidence, the outgoing phonon distribution no longer follows a simple cosine law (Fig. 6.8). We will first attempt to use the inversion technique to find the correct distribution of scattered phonons; after encountering difficulties, we will turn to the rejection technique.

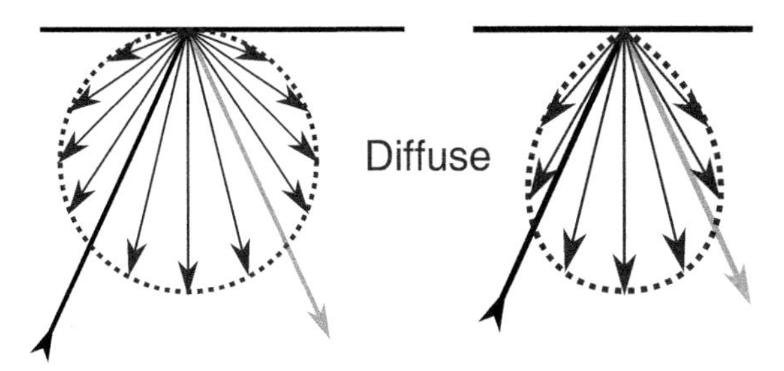


**Figure 6.8** Polar plots of the angular distribution of the outgoing phonon momentum upon diffuse scattering from a surface with (left) a constant specularity parameter and (right) Soffer's momentum-dependent specularity parameter (Eq. 6.24) [21]. Soffer's momentum-dependent specularity parameter decreases the chance of diffuse scattering as $\theta$ (the angle between the phonon wave vector and the surface normal) increases. To satisfy detailed balance, the distribution of outgoing phonons must match the probability of diffuse scattering. The result is a teardrop-shaped distribution, which suppresses scattering at large $\theta$ relative to the constant specularity parameter model.

The angular distribution of phonons incident on the surface is still proportional to $\cos\theta$, and the probability of diffuse scattering is $1 - p(\theta, \phi)$, so the distribution of incoming phonons that will be diffusely scattered is

$$\begin{aligned} f(\theta, \phi) &= C\cos\theta\,(1 - p(\theta, \phi)) \\ &= C\cos\theta\,(1 - e^{-(2\sigma|\mathbf{q}|\cos\theta)^2}), \end{aligned} \tag{6.25}$$

where $C$ is a normalization constant given by

$$\begin{aligned} 1 &= \int_0^{\pi/2}\int_0^{2\pi} f(\theta',\phi')\sin\theta' d\theta' d\phi', \\ C^{-1} &= 2\pi\left(1 - \frac{1-e^{-4\sigma^2 q^2}}{4\sigma^2 q^2}\right). \end{aligned} \tag{6.26}$$

Now that $C$ is known, we can find the CDF as

$$\begin{aligned} F(\theta) &= \int_0^{\theta}\int_0^{2\pi} f(\theta',\phi')\sin\theta' d\theta' d\phi' \\ &= \frac{\dfrac{e^{-4\sigma^2 q^2} - e^{-4\sigma^2 q^2\cos^2(\theta)}}{4\sigma^2 q^2} + \sin^2(\theta)}{1 - \dfrac{1-e^{-4\sigma^2 q^2}}{4\sigma^2 q^2}}. \end{aligned} \tag{6.27}$$

We have been able to do the integration step analytically, but we cannot do the inversion step analytically. So,

$$r_\theta = \frac{\dfrac{e^{-4\sigma^2 q^2} - e^{-4\sigma^2 q^2 \cos^2(\theta)}}{4\sigma^2 q^2} + \sin^2(\theta)}{1 - \dfrac{1 - e^{-4\sigma^2 q^2}}{4\sigma^2 q^2}} \tag{6.28}$$

will have to be solved numerically, for example, with Newton's method (which requires computing $\frac{dF}{d\theta}$) or the bisection method.

The other option is to use the rejection technique. This is particularly appealing because we already have a good distribution function $g(\theta, \phi) \geq f(\theta, \phi)$, namely $g(\theta, \phi) = \cos\theta$, which is the distribution we use for the constant specularity parameter model.[d] The rejection method to find the outgoing angle $\theta'$ of a scattered phonon is then:

(1) Generate a random variate $\theta' = \arcsin\left(\sqrt{r_\theta}\right)$ where $r_\theta$ is uniformly distributed in [0, 1].
(2) Generate a random variate $y$ that is uniformly distributed in $[0, \cos\theta']$.
(3) If $y < f(\theta', \phi) = \cos\theta' \left(1 - e^{-(2\sigma|\mathbf{q}| \cos\theta')^2}\right)$, then use $\theta'$ as your random variate. Otherwise, return to step 1.

Because we have a good $g(\theta, \phi)$, and because the inversion method will require an iterative solution, the rejection method is both faster and simpler to implement.

## 6.6 Contacts

In the PMC transport simulation, we need contacts at the ends of the simulation domain to act as fixed-temperature phonon reservoirs. These reservoirs must inject new phonons into the simulation domain and absorb the phonons that leave it. There are two basic methods to implement contacts: Either the boundaries of the simulation domain mimic reservoirs outside the simulation domain,

[d]We could also use a constant distribution function for $g(\theta, \phi)$ as described in Section 6.2.2, but that would result in reduced performance.

or parts of the simulation domain get turned into reservoirs. We will call the former a boundary contact and the latter an internal contact.

Both approaches can make use of existing code, because creating new phonons has algorithmically much in common with scattering existing phonons. Internal contacts can reuse the code for internal scattering mechanisms, and boundary contacts can reuse the code for diffuse surface scattering. All PMC simulations will implement internal scattering mechanisms, so it is always straightforward to implement internal contacts. In contrast, not all PMC simulations will implement diffuse surface scattering, so it may take extra work to implement boundary contacts. We will see that boundary contacts require fewer new phonons to be generated each time step, which increases the performance of the simulation.

### 6.6.1 3D Internal Contacts

Internal contacts are relatively simple to implement because the approach almost exclusively reuses code from other parts of the simulation. To implement internal contacts in 3D, one simply deletes all the phonons in a volume at the end of the simulation domain and then fills the volume with new phonons drawn from the equilibrium distribution (Section 6.4.2). The only size requirement is that the volume must be large enough that no phonons can traverse the volume from end to end in one time step.

There are two equivalent ways to fill the volume: with a certain number of phonons (fill by number) and by adding phonons until the cell has the correct energy (fill by energy). It is simpler to implement code that fills by number, but the simulation will already need code to fill volumes by energy, in which case filling by energy can simply reuse existing code. We will discuss how to fill by energy in Section 6.7, so we explain how to fill by number here. (Boundary contacts will also use number rather than energy.)

The average number of phonons in branch b in the volume at a temperature $T$ is[e]

$$\mathcal{N}_{\mathrm{b}}(T)=V\int\frac{d^3\mathbf{q}}{(2\pi)^3}\langle n_{\mathrm{BE}}(\omega_{\mathrm{b},q},T)\rangle, \tag{6.29}$$

[e]Eq. 6.29 is similar to Eq. 6.9 except that that the latter uses the DOS instead of an integral over reciprocal space.

where the integral overall allowed is $\mathbf{q}$ (the Brillouin zone in the case of a full dispersion relation). The expression for the average energy in the volume is the same except that a factor of $\hbar\omega_i(\mathbf{q})$ is included in the integral.

The above expression can be simplified for an isotropic dispersion relation:

$$\mathcal{N}_\mathrm{b} = V\int_0^{2\pi}\int_0^{\pi}\int_0^{q_\mathrm{max}} \langle n_\mathrm{BE}(\omega_{\mathrm{b},\mathbf{q}},T)\rangle \frac{q^2 \sin\theta dq d\theta d\phi}{(2\pi)^3},$$
$$= \frac{V}{2\pi^2}\int_0^{q_\mathrm{max}} \langle n_\mathrm{BE}(\omega_{\mathrm{b},q},T)\rangle q^2 dq, \tag{6.30}$$

where $q = |\mathbf{q}|$.

$\mathcal{N}_\mathrm{b}$ will not be an integer in general. So, add $\lfloor \mathcal{N}_\mathrm{b} \rfloor$ phonons and then randomly, with a probability of $\mathcal{N}_\mathrm{b} - \lfloor \mathcal{N}_\mathrm{b} \rfloor$, add one more phonon. For example, if $\mathcal{N}_\mathrm{b} = 1192.63$, then always add 1192 phonons and add an additional phonon 63% of the time. In this case, it might seem that the additional phonon probability can be done without, but incorrectly generating even one excess phonon per time step per cell may lead to a failure of the simulation (see Section 6.7).

### 6.6.2 3D Boundary Contacts

The advantage of boundary contacts is that instead of having to generate all the phonons inside a volume every time step, you only need to generate the phonons that would have drifted into the simulation domain from a reservoir outside the simulation domain. This effectively makes a boundary contact a blackbody that emits phonons into the simulation domain, so we can use results for the theory of blackbody radiation [20]. Boundary contacts have much in common with diffuse scattering (Section 6.5).

The expectation number of phonons of branch b and wave vector $\mathbf{q}$ passing through a surface per unit area and time is

$$n_\mathrm{b}(\mathbf{q}, T) = |\mathbf{v}_{\mathrm{g,b}}(\mathbf{q})| \cos\theta \, \langle n_\mathrm{BE}(\omega_{\mathrm{b},\mathbf{q}},T)\rangle \frac{d^3\mathbf{q}}{(2\pi)^3}, \tag{6.31}$$

where $\theta$ is the angle between the group velocity $\mathbf{v}_{\mathrm{g,b}}(\mathbf{q})$ and the surface normal. Because we are only interested in the number of phonons entering the domain, we only consider $\theta \in [0, \pi/2]$.

We now turn to the case of an isotropic dispersion relation with $q_{\max} \geq |\mathbf{q}|$ as an upper bound on the wave-vector magnitude. Then $n_b$ $(\mathbf{q}, T)$ can be simplified to [9]

$$n_b(q,T) = \int_0^{\pi/2}\int_0^{2\pi} \mathbf{v}_{g,b}(q)\cos\theta\langle n_{BE}(\omega_{b,\mathbf{q}},T)\rangle \frac{q^2 \sin\theta dq d\theta d\phi}{(2\pi)^3},$$

$$= \mathbf{v}_{g,b}(q)\,\langle n_{BE}(\omega_{b,\mathbf{q}},T)\rangle \frac{q^2}{8\pi^2}\,dq. \quad (6.32)$$

Then the total number of phonons entering the simulation domain due to a blackbody of area $A$ in a period of time $\Delta t$ is

$$\mathcal{N}_{bc}^{3D}(T) = A\Delta t \sum_b \int_0^{q_{\max}} v_{g,b}(q)\cos\theta\langle n_{BE}(\omega_{b,\mathbf{q}},T)\rangle \frac{q^2}{8\pi^2}\,dq. \quad (6.33)$$

The cosine in Eq. 6.31 means that phonons entering the simulation will have the angular distribution from Lambert's cosine law, so we can generate random directions for the incoming phonons using Eq. 6.4. In principle, the phonons from the boundary contact will enter the simulation domains at different times, but in practice, all the phonons can be added to the simulation domain at the same time because $\Delta t$ is small.

Then, the overall procedure is to calculate $\mathcal{N}_{bc}^{3D}$ $(T_h)$ and $\mathcal{N}_{bc}^{3D}$ $(T_c)$ for the hot and cold contacts, which are at temperatures $T_h$ and $T_c$, respectively. Then at each time step, any phonons that drift outside the simulation domain are deleted and $\mathcal{N}_{bc}^{3D}$ $(T_h)$ and $\mathcal{N}_{bc}^{3D}$ $(T_c)$ equilibrium phonons are created (Section 6.4.2) at the two contacts, with angular distributions given by Lambert's cosine law (Eq. 6.4).

### 6.6.3 2D Contacts

2D internal contacts are essentially the same as their 3D counterparts, except we use the cell area instead of the volume and the integration is over a 2D wave vector $\mathbf{q}$. When using internal contacts, it does not matter what kind of dispersion we use, because we just initialize the cell to the reservoir energy.

2D boundary contacts need more care in implementation. In 2D, to mimic a reservoir from the outside, we need to delete phonons that drift outside of the end-cell boundary (a line) and inject phonons into the cell as if they come from an outside reservoir fixed at a temperature $T$. For isotropic dispersion, it is easy to work out the expectation number of phonon to be injected per unit length, per branch, per unit wave number, and per unit time as[f]

$$n_{\mathrm{b}}(q,T) = \frac{1}{4\pi}\langle n_{\mathrm{BE}}(\omega_{c,j},T)\rangle \frac{q}{2\pi^2}\mathbf{v}_{\mathrm{g,b}}(q), \tag{6.34}$$

where $\mathbf{v}_{\mathrm{g,b}}(q)$ is the group velocity for phonon with wave number $q$ and branch b. The total expectation number per time step is then

$$\mathcal{N}_{\mathrm{bc}}^{\mathrm{2D}} = H\Delta t \sum_{\mathrm{b}} \int_0^{q_{\max}} n_{\mathrm{b}}(q)\, dq, \tag{6.35}$$

where $\Delta t$ is the time step and $H$ is the height of the simulation domain. Like in 3D, we only generate $\lfloor \mathcal{N}_{\mathrm{bc}}^{\mathrm{2D}} \rfloor$ phonons each step and add an additional phonon with a chance of $\left(\mathcal{N}_{\mathrm{bc}}^{\mathrm{2D}} - \lfloor \mathcal{N}_{\mathrm{bc}}^{\mathrm{2D}} \rfloor\right)$. The CDF of angle in 2D is

$$F(\theta) = \frac{\int_0^{\theta} \cos\theta' d\theta'}{\int_0^{\pi} \cos\theta' d\theta'} = \sin\theta. \tag{6.36}$$

The $\cos\theta'$ term comes from the fact that phonons come in at a rate proportional to the group velocity perpendicular to the boundary. Then the inversion method can be used to choose the angle as $\theta = \arcsin(r_\theta)$, just like for 2D diffuse scattering (Eq. 6.23). When $H$ is small, which is typically the case for PMC for quasi-1D graphene ribbons, the expectation number of injected particles is too small (<10) to represent the distribution, even though the distribution would be correct if we drew a large number of phonons. As a result, the 2D boundary contacts have stability issues and the simulation may take a longer time to converge than the internal contact implementation.

## 6.7 Energy Conservation

If we use the rejection technique to generate a phonon in a cell for PMC simulation, the phonon energy is independent of how much

[f]Compare with the 3D equivalent in Eq. 6.32

energy is already in the cell. As a result, it is impossible to infuse the cell with exactly the expected energy. It is tempting to force the last phonon's energy to achieve exact energy conservation. However, by doing so, we are introducing excess phonons with energies smaller than average, that is, more of them than the equilibrium distribution would predict. Owing to the low energies, these phonons have large group velocities and small scattering rates, so they carry their energy out of the simulation domain almost ballistically and lead to energy depletion inside the structure. Even generating a single phonon from the wrong distribution in every cell once a time step can add up and cause the simulation to fail dramatically.

To respect the phonon energy distribution, we allow the energy inside a cell to be uncertain up to half the maximal energy of a single phonon: we consider the desired cell energy to be reached with satisfactory accuracy if the cell energy falls within $\left[E-\frac{\hbar\omega_{\mathrm{max}}}{2}, E+\frac{\hbar\omega_{\mathrm{max}}}{2}\right]$, where $E$ is the exact desired cell energy and $\hbar\omega_{\mathrm{max}}$ is the maximum energy carried by a phonon in the material.

Furthermore, the treatment of inelastic scattering does not conserve energy precisely at each time step but only in an average sense [12]. To better conserve the energy after scattering, each cell is reinitialized after each time step by adding phonons to or deleting them from the equilibrium distribution. The accumulation of offset energy might cause a problem, so we record the offset energy $E_{\mathrm{offset}} = E - E_{\mathrm{actual}}$ at each step and add it to the desired energy at the next reinitialization.

During drift, phonons might cross cell boundaries and, therefore, change the total energy in each cell. The energy of the phonons that drifted in (or drifted out) has to be added (or subtracted) before entering the scattering routine. As a result, we record the cell energy just after the phonons finish drift in a prescattering array, $E_{\mathrm{prescat}}$. Together with the array $E_{\mathrm{i,offset}}$, which records the offset energy from our last attempt at enforcing energy conservation, we calculate the desired energy after scattering as $E_{\mathrm{d}} = E_{\mathrm{prescat}} + E_{\mathrm{offset}}$. Then we can scatter the phonons and calculate the actual cell energy after scattering, stored in an after-scattering array, $E_{i,\mathrm{afterscat}}$. Again, we only enforce that the cell energy $E$ gets into the range $\left[E_{\mathrm{d}}-\frac{\hbar\omega_{\mathrm{max}}}{2}, E_{\mathrm{d}}+\frac{\hbar\omega_{\mathrm{max}}}{2}\right]$, where $E_{\mathrm{d}}$ is the desired energy. The initial cell energy in the reinitialization process is $E_i = E_{i,\mathrm{afterscat}}$. For each cell $i$, we compare $E_i$ and $E_{i,\mathrm{d}}$.

(1) If $E_i \in \left[E_{i,\mathrm{d}} - \frac{\hbar\omega_{\max}}{2}, E_{i,\mathrm{d}} + \frac{\hbar\omega_{\max}}{2}\right]$, we consider this cell good and move on to the next one.

(2) If $E_i < E_{i,\mathrm{d}} - \frac{\hbar\omega_{\max}}{2}$, we generate a phonon from the equilibrium distribution and add it to a random place in the cell. The new cell energy after addition is $E_i^{\mathrm{new}} = E_i^{\mathrm{old}} + \hbar\omega_0$, where $\hbar\omega_0$ is the energy carried by the added phonon. Keep adding phonons until the final $E_i$ falls in the appropriate range.

(3) If $E_i > E_{i,\mathrm{d}} + \frac{\hbar\omega_{\max}}{2}$, we would randomly choose one phonon in the cell and delete it.[g] After the deletion, the new energy in the cell is $E_i^{\mathrm{new}} = E_i^{\mathrm{old}} - \hbar\omega_0$, where $\hbar\omega_0$ is the energy carried by the deleted phonon. Then we compare $E_i^{\mathrm{new}}$ and $E_{i,\mathrm{d}}$ again and keep this random deletion until $E_i^{\mathrm{new}}$ falls in the desired range.

After all cells have energies in the desired range, we record the new offset energy as $E_{\mathrm{offset}} = E_{\mathrm{d}} - E_i$ and use it in the next reinitialization process (after the next time step).

## 6.8 Conclusion

PMC is a versatile stochastic technique for solving the Boltzmann equation for phonons in structures that can have real-space roughness and experimentally relevant sizes. We presented the relative merits of inversion versus rejection techniques for generating random variates to represent the random variables with nonuniform distributions, which are relevant in thermal transport: generating the attributes for phonons in equilibrium with full and

[g]It is very important that the phonon be chosen randomly. For example, simply deleting the oldest phonon will slowly skew the phonon distribution because the oldest phonon is likely to be a phonon with a low scattering rate. Depending on the data structure used for storing phonons, deleting algorithms may vary. One can use a random number to choose the index of phonon to be deleted if the structure supports easy random access. For structures like the linked list, one can use the Fisher–Yates shuffle algorithm to generate a random permutation of the array and delete in a sequence.

isotropic dispersions, randomizing outgoing momentum upon diffuse boundary scattering, implementing contacts (boundary and internal), and conserving energy in the simulation. We also identified common themes in phonon generation and scattering that are helpful for reusing code in the simulation (generating thermal phonon attributes versus internal contacts; diffuse surface scattering versus boundary contacts). We hope these examples will inform the reader about both the mechanics of random-variate generation and choosing a good approach for whatever problem is at hand, and aid in the more widespread use of PMC for thermal transport simulation.

## Acknowledgments

The authors gratefully acknowledge support by the US Department of Energy, Office of Basic Energy Sciences, Division of Materials Sciences and Engineering, under Award DE-SC0008712. This work was performed using the compute resources and assistance of the UW–Madison Center for High Throughput Computing (CHTC) in the Department of Computer Sciences.

## References

1. Ziman, J. M. (1960). *Electrons and Phonons: The Theory of Transport Phenomena in Solids*, Clarendon Press.
2. Nika, D. L., and Balandin, A. A. (2012). Two-dimensional phonon transport in graphene, *J. Phys. Condens. Matter*, **24**(23), p. 233203, doi:10.1088/0953-8984/24/23/233203, http://www.ncbi.nlm.nih.gov/pubmed/22562955.
3. Broido, D. A., Malorny, M., Birner, G., Mingo, N., and Stewart, D. A. (2007). Intrinsic lattice thermal conductivity of semiconductors from first principles, *Appl. Phys. Lett.*, **91**, p. 231922.
4. Aksamija, Z., and Knezevic, I. (2010). Anisotropy and boundary scattering in the lattice thermal conductivity of silicon nanomembranes, *Phys. Rev. B*, **82**, p. 045319.
5. Peterson, R. B. (1994). Direct simulation of phonon-mediated heat transfer in a Debye crystal, *J. Heat Transfer*, **116**(4), pp. 815–822, doi:10.1115/1.2911452, http://heattransfer.asmedigitalcollection.asme.org/article.aspx?articleid=1441842.
6. Mazumder, S., and Majumdar, A. (2001). Monte Carlo study of phonon transport in solid thin films including dispersion and polarization, *J. Heat*

*Transfer*, **123**(4), p. 749, doi:10.1115/1.1377018, http://heattransfer.asmedigitalcollection.asme.org/article.aspx?articleid=1445087.

7. Lacroix, D., Joulain, K., and Lemonnier, D. (2005). Monte Carlo transient phonon transport in silicon and germanium at nanoscales, *Phys. Rev. B*, **72**, p. 064305, doi:10.1103/PhysRevB.72.064305, http://link.aps.org/doi/10.1103/PhysRevB.72.064305.
8. Lacroix, D., Joulain, K., Terris, D., and Lemonnier, D. (2006). Monte Carlo simulation of phonon confinement in silicon nanostructures: application to the determination of the thermal conductivity of silicon nanowires, *Appl. Phys. Lett.*, **89**(10), p. 103104, doi:10.1063/1.2345598, http://scitation.aip.org/content/aip/journal/apl/89/10/10.1063/1.2345598.
9. Randrianalisoa, J., and Baillis, D. (2008). Monte Carlo simulation of steady-state microscale phonon heat transport, *J. Heat Transfer*, **130**, p. 072404, doi:10.1115/1.2897925, http://heattransfer.asmedigitalcollection.asme.org/article.aspx?articleid=1449183.
10. Ramayya, E. B., Maurer, L. N., Davoody, A. H., and Knezevic, I. (2012). Thermoelectric properties of ultrathin silicon nanowires, *Phys. Rev. B*, **86**, p. 115328, doi:10.1103/PhysRevB.86.115328, http://link.aps.org/doi/10.1103/PhysRevB.86.115328.
11. Bera, C. (2012). Monte Carlo simulation of thermal conductivity of Si nanowire: an investigation on the phonon confinement effect on the thermal transport, *J. Appl. Phys.*, **112**, p. 074323.
12. Mei, S., Maurer, L. N., Aksamija, Z., and Knezevic, I. (2014). Full-dispersion Monte Carlo simulation of phonon transport in micron-sized graphene nanoribbons, *J. Appl. Phys.*, **116**(16), p. 164307, doi:10.1063/1.4899235, http://scitation.aip.org/content/aip/journal/jap/116/16/10.1063/1.4899235.
13. Péraud, J.-P. M., Landon, C., and Hadjiconstantinou, N. (2014). Monte Carlo methods for solving the Boltzmann transport equation, *Annu. Rev. Heat Transfer*, **17**, pp. 205–265.
14. Maurer, L. N., Aksamija, Z., Ramayya, E. B., Davoody, A. H., and Knezevic, I. (2015). Universal features of phonon transport in nanowires with correlated surface roughness, *Appl. Phys. Lett.*, **106**(13), p. 133108, doi:10.1063/1.4916962, http://scitation.aip.org/content/aip/journal/apl/106/13/10.1063/1.4916962.
15. Ramiere, A., Volz, S., and Amrit, J. (2016). Geometrical tuning of thermal phonon spectrum in nanoribbons, *J. Phys. D: Appl. Phys.*, **49**(11), p. 115306, http://stacks.iop.org/0022-3727/49/i=11/a=115306.

16. Glassbrenner, C. J., and Slack, G. A. (1964). Thermal conductivity of silicon and germanium from 3° K to the melting point, *Phys. Rev.*, **134**, p. A1058.

17. Esfarjani, K., Chen, G., and Stokes, H. T. (2011). Heat transport in silicon from first-principles calculations, *Phys. Rev. B*, **84**, p. 085204.

18. Mei, S., and Knezevic, I. (2015). Thermal conductivity of III-V semiconductor superlattices, *J. Appl. Phys.*, **118**(17), p. 175101, doi:http://dx.doi.org/10.1063/1.4935142, http://scitation.aip.org/content/aip/journal/jap/118/17/10.1063/1.4935142.

19. Devroye, L. (1986). *Non-Uniform Random Variate Generation*, Springer-Verlag.

20. Modest, M. F. (2003). *Radiative Heat Transfer*, Academic Press.

21. Soffer, S. B. (1967). Statistical model for the size effect in electrical conduction *J. Appl. Phys.*, **38**(4), pp. 1710–1715.

22. Lim, J., Hippalgaonkar, K., Andrews, S. C., Majumdar, A., and Yang, P. (2012). Quantifying surface roughness effects on phonon transport in silicon nanowires, *Nano Lett.*, **12**(5), pp. 2475–2482, doi:10.1021/nl3005868.

23. Berman, R., Simon, F. E., and Ziman, J. M. (1953). The thermal conductivity of diamond at low temperatures, *Proc. R. Soc. A*, **220**(1141), pp. 171–183, doi:10.1098/rspa.1953.0180, http://rspa.royalsocietypublishing.org/content/220/1141/171.

24. Berman, R., Foster, E. L., and Ziman, J. M. (1955). Thermal conduction in artificial sapphire crystals at low temperatures. I. Nearly perfect crystals, *Proc. R. Soc. A*, **231**(1184), pp. 130–144, doi:10.1098/rspa.1955.0161, http://rspa.royalsocietypublishing.org/content/231/1184/130.

25. Northrop, G. A., and Wolfe, J. P. (1984). Phonon reflection imaging: a determination of specular versus diffuse boundary scattering, *Phys. Rev. Lett.*, **52**, pp. 2156–2159, doi:10.1103/PhysRevLett.52.2156, http://link.aps.org/doi/10.1103/PhysRevLett.52.2156.

26. Couchman, L., Ott, E., and Antonsen, T. M. (1992). Quantum chaos in systems with ray splitting, *Phys. Rev. A*, 46, pp. 6193–6210, doi:10.1103/PhysRevA.46.6193, http://link.aps.org/doi/10.1103/PhysRevA.46.6193.

# Chapter 7

# Hybrid Photovoltaic-Thermoelectric Solar Cells: State of the Art and Challenges

**Bruno Lorenzi and Dario Narducci**
*University of Milano Bicocca, Department of Materials Science, via R. Cozzi 55, I-20125 Milan, Italy*
dario.narducci@unimib.it

## 7.1 Introduction

Concerns about sufficient availability of fossil sources in the near future to satisfy the increasing demand for energy, along with an increasing consciousness about global warming and emission-related climate change, have further pushed research and technology development of the so-called renewable energy sources. Solar energy is by far the most widely used of such sources. Photovoltaic (PV) cells have actually undergone an impressive efficiency enhancement over the last decades, stepping up from ~13% in 1977 (single-junction, nonconcentrated single-crystalline solar cells) to

*Nanophononics: Thermal Generation, Transport, and Conversion at the Nanoscale*
Edited by Zlatan Aksamija

ISBN 978-981-4774-41-3 (Hardcover), 978-1-315-10822-3 (eBook)
www.panstanford.com

the current record efficiency of 44% in concentrated, three-junction lattice-matched solar cells [1]. At the same time, solar energy costs have dropped from 77 USD/W (1977) to 0.30 USD/W (2015) [2]. Currently, PV research is addressing two major avenues, namely that of further enhancing PV efficiency and that of developing cells based on noncritical raw materials. On the former side, spectral matching is pursued. Actually, as will be shown in greater detail in Section 7.4, optimal photon energy conversion is achieved only for photons with energy very close to the PV material energy bandgap $E_g$. Thus, enhanced efficiency may be obtained either by stacking materials with different $E_g$ in multijunction cells or by reshaping the solar spectrum so as to increase the spectral fraction that may be more efficiently converted into electric energy by a given set of PV materials. On the latter side, instead, new PV materials have surfaced over the last years, including solid solutions such as the CIGS ($CuIn_xGa_{1-x}Se_2$) [3] and the CZTS ($Cu_2ZnSnS_{4-x}Se_x$) family [4].

In principle, both aims may be achieved also by hybridizing PV cells with non-PV conversion devices able to recover the heat released by nonoptimized PVs. This chapter actually discusses how thermoelectric generators (TEGs) harvesting the heat released by PV cells may impact efficient solar energy conversion. Focus will be on terrestrial applications and on inorganic single-junction PV cells. After a primer on thermoelectricity, the state of the art on solar TEG strategies will be discussed, including all-thermoelectric (TE) solar conversion and thermodynamic cogeneration—moving then to the theory behind photovoltaic-thermoelectric (PV-TE) hybridization. After discussing the fundamentals of PV physics and technology, the available heat released by PV cells will be analyzed, also in view of the dependence of PV efficiency upon the cell temperature. It will be shown that tandem PV-TE devices may lead to increased power densities when an intermediate layer absorbing the under-the-gap fraction of the solar spectrum is implemented. However, effective heat dissipation (TE thermal matching) will be shown to be a key issue for proper PV-TE coupling. The layout of the TEG will be demonstrated to need redesigning to make tandem cells effective. As an alternative, optical or thermal concentrations are needed.

## 7.2 A Primer on Thermoelectricity

TE effects are an instance of a larger class of physical phenomena of cross-coupling between thermodynamic fluxes and forces, properly accounted for in linear nonequilibrium thermodynamics [5]. Heat fluxes $\vec{J}_h$ and electrical current densities $\vec{J}_q$ depend upon the applied electrical field $\vec{F}$ and the temperature gradient $\nabla T$ as

$$\begin{aligned} \vec{J}_h &= kT^2\nabla\left(\frac{1}{T}\right) \quad +\rho^{-1}\alpha T^2\left(\frac{\vec{F}}{T}\right) \\ \vec{J}_q &= \rho^{-1}\alpha T^2\nabla\left(\frac{1}{T}\right)+\rho^{-1}T\left(\frac{\vec{F}}{T}\right) \end{aligned}, \qquad (7.1)$$

where $\alpha$ is the Seebeck coefficient, $\rho$ is the electrical resistivity, and $\kappa$ is the thermal conductivity. All quantities are assumed to be independent of the temperature, a sensible approximation for small temperature gradients. Therefore, the application of a temperature gradient generates an open-circuit TE voltage (Seebeck effect) while in a shorted circuit an electrical current forces heat to flow from the cold to the hot sink (Peltier effect).

TEGs are heat engines using thermoelectricity to partially convert heat into electric energy. In their most common configuration TEGs are electrical circuits of alternated p- and n-type semiconductors connected in series, forming a parallel thermal circuit with the hot sink (at a temperature $T_H$) and the cold sink (at a temperature $T_C$). Since the TE voltage occurs in a short-circuited configuration, an analysis of the generator efficiency has to account not only for the Seebeck effect (and for the pertinent electric current) but also for the heat flowing from the cold to the heat sink due to the Peltier effect, associated with the nonzero electric current flowing through the circuit. Following Altenkirch's and Ioffe's arguments [6], in a two-leg circuit made of TE materials of length $d$ and sections $A_1$ and $A_2$, with electrical resistance $r_1$ and $r_2$ and thermal conductance $k_1$ and $k_2$, operating between two heat sinks at $T_H$ and $T_C$ ($<T_H$), the full heat balance must account for the heat flowing through the legs by thermal conduction $J_{th} = (k_1 + k_2)\Delta T$ (where $\Delta T = T_H - T_C$), the Peltier heating (cooling) due to the current $i$ flowing through the circuit $J_H = \alpha_H T_H i$ and $J_C = \alpha_C T_C i$, and the heat generated by the Joule effect, $J_J = i^2(r_1 + r_2)$.

The electric power $W$ generated by the circuit on an electric load $R$ is simply given by $W = i^2R$, where the current flowing through the circuit is controlled by the TE voltage $Uth = (|\alpha_1| + |\alpha_2|)\Delta T$. Thus, defining $\alpha \equiv |\alpha_1| + |\alpha_2|$, $k \equiv k_1 + k_2$ and $r \equiv r_1 + r_2$, the output power reads

$$W = \left(\frac{\alpha \Delta T}{R+r}\right)^2 R \tag{7.2}$$

Thus the conversion yield $\varphi$ reads

$$\phi = \frac{T_H - T_C}{T_H} \frac{m/(m+1)}{1+\frac{kr}{\alpha^2}\frac{m+1}{T_H} - \frac{1}{2}\frac{\Delta T}{T_H(m+1)}}, \tag{7.3}$$

where $m \equiv R/r$.

The actual maximum thermodynamic conversion efficiency can be obtained by maximizing $\varphi$ twice. Firstly, one may minimize the heat flow dissipated by direct thermal conduction by minimizing $kr$. In turn this implies an optimization over the leg cross sections. Simple algebra shows that

$$\max_{kr} \phi \equiv \tilde{\phi} = \eta_{\text{Carnot}} \frac{m/(m+1)}{1+\frac{m+1}{ZT_{\text{H}}} - \frac{1}{2}\frac{\Delta T}{T_{\text{H}}(m+1)}} \tag{7.4}$$

where $\eta_{\text{Carnot}} = (T_{\text{H}} - T_{\text{C}})/T_{\text{H}}$ and a thermoelectric figure of merit of the paired legs $Z_{12}T$ is defined as

$$Z_{12}T = \frac{\alpha^2 T}{(\sqrt{k_1\rho_1} + \sqrt{k_2\rho_2})^2} \tag{7.5}$$

where $\rho_i$ and $\kappa_i$ ($i$ = 1, 2) are the electrical resistivity and the thermal conductivity of the two legs, respectively. By extension, it is commonplace to define a corresponding TE figure of merit for a single material as

$$ZT = \frac{\alpha^2 T}{k\rho} \tag{7.6}$$

A second optimization of $\tilde{\phi}$ (over $m$) is needed to obtain the actual TE efficiency. The maximum output power is delivered by the TEG when the load is matched to the generator, that is, for $m = 1$. This leads to

$$\eta_{\mathrm{w}} = \eta_{\mathrm{Carnot}} \frac{1/2}{1+\dfrac{2}{Z_{12}T_{\mathrm{H}}} - \dfrac{DT}{4T_{\mathrm{H}}}} \tag{7.7}$$

The maximum efficiency is obtained instead for $\partial\tilde{\phi}/\partial m = 0$. Thus, $m = \sqrt{1+Z_{12}\bar{T}}$ (where $\bar{T} = (T_{\mathrm{H}} + T_{\mathrm{C}})/2$), so

$$\eta_{\mathrm{TE}} = \eta_{\mathrm{Carnot}} \frac{\sqrt{1+Z_{12}\bar{T}} - 1}{\sqrt{1+Z_{12}\bar{T} + T_{\mathrm{C}}/T_{\mathrm{H}}}}. \tag{7.8}$$

It is worthwhile to remark that the highest power output for a given TEG is obtained when $r = R$ with an efficiency $\eta_{\mathrm{w}} < \eta_{\mathrm{TE}}$. Actually, however, $\eta_{\mathrm{w}}$ and $\eta_{\mathrm{TE}}$ significantly differ from each other only at high temperatures and for high $Z$ values.

## 7.3 Strategies of Thermoelectric Solar Energy Conversion

Solar TE systems have been the object of research since Telkes's seminal work on TE conversion of solar energy. Using Zn-Sb and Bi-Sb elements Telkes attained a conversion efficiency of 3.35% using solar concentration [7]. In the same year a 4% efficiency was claimed for the first lithium-silicon PV cell [8]. However, since then PV conversion began its rush toward two-digit efficiencies while no comparable improvement of TE efficiency occurred, so research on TE solar conversion stopped.

As shown in Eq. 7.6, large TE figures of merit actually require large $|\alpha|$'s and small $\rho$'s along with small $\kappa$'s. In almost all standard materials large $|\alpha|$ values occur in electrical insulators while Wiedemann–Franz law binds the value of $1/\rho_i\kappa_i$. Nanotechnology could actually break the interdependency among transport coefficients [9, 10], and TE materials could overcome the $ZT = 1$ threshold, reviving the interest for TE applications. As of today, nanostructured materials have overcome $ZT$ values of 2 [11], also revamping materials (e.g., silicon) that for long times had been considered of little relevance for thermoelectricity [12–14]. This has enabled the development of commercial TEGs with efficiencies larger than 10% [15].

In the use of thermoelectricity for solar energy conversion three lines of action have been considered:

- Solar thermoelectric generators (STEGs), converting solar power into heat and then converting it into electricity by using TEGs
- Hybrid cogenerative solar thermoelectric generators (HCG-STEGs), integrating STEGs with a second harvester used to recover the heat released by TEGs to generate hot water
- Hybrid thermoelectric-photovoltaic generators (HTEPVGs), using TEGs in a tandem configuration with a PV cell

### 7.3.1 Solar Thermoelectric Generators

Modern STEGs take advantage of nanostructured TE materials to directly convert solar power into heat and then into electric power, also replacing optical concentration with thermal concentration. In principle, the heat generated at the front of the STEG may be concentrated to flow through an arbitrarily small TEG, with the advantage of saving TE material costs and raising the hot-side temperature. However, thermal concentration is challenging because of the hurdles of driving heat currents. As an example, the best STEG, developed at MIT in 2011 [16], operates under vacuum, and special care is taken to limit radiative emission from surfaces. An efficiency of 4.6% could then be achieved, seven to eight times larger than any previously reported yield in likely devices [17–21]. Manifestly enough, efficiency remains less than one-tenth of the best PV efficiency. Thus, while the approach remains of remarkable interest, it cannot directly compete nowadays with PV conversion rates.

### 7.3.2 Hybrid Cogenerative Solar Thermoelectric Generators

In view of the moderate efficiency of STEGs, coupling with fluid heaters may mitigate low conversion rates. A number of HCG-STEGs has been studied and deployed [22–25]. Solar power is converted into heat by suitable optical absorbers also acting as the hot sink of the TEG. Heat dissipated from the TEG cold side is then input into a thermal exchanger to warm up a suitable fluid. Although electrical efficiencies are usually very low (on the order of some percentage points), a HCG-STEG should be probably better seen as a thermal solar plant with a TEG added on. Compared to standard hybrid cogenerative photovoltaic generators (HCG-PVG), HCG-STEGs also generate a (small) amount of electric power, which complements thermal solar conversion. A remarkable example of a large HCG-STEG plant not using solar concentrators was reported by Zhang et al. [26]. Heat pipes are used to transfer heat to the hot side of the TEG, while cold water refrigerates its cold side, leaving the TEG at a temperature of about 55°C. An average electric power of 63 W is reported, while 100 *l*/h of warm water is generated.

### 7.3.3 Hybrid Thermoelectric-Photovoltaic Generators

HTEPVGs face the issue of reusing heat released by the backside of PV cells by implementing a TEG in series to the PV cell itself. They share some of the critical issues of HCG-PVGs as they also require the backside of the PV cell to be kept at a relatively high temperature, while, differently from STEGs, they take advantage of the large PV conversion efficiency for that part of the solar spectrum that PV may actually convert.

As mentioned, PV cells efficiently convert only photons at the frequency corresponding to the energy gap of PV absorbing component(s). Photons with smaller energy are actually not absorbed, while photons with larger energy partially convert their energy into heat, thus setting the cell temperature. Therefore, in principle thermoelectrics may be used to recover and convert such heat into electricity. A number of theoretical and experimental papers have been published on these systems. Theoretical studies include evaluations of the hybrid thermoelectric-photovoltaic (HTEPV) power output and efficiency [27–29], along with electrical optimization models [30,31], in a host of configurations. Experimental works analyzed instead direct coupling with dye-sensitized solar cells (DSSCs) [32–34], electrical optimization between silicon solar cell and TEG modules [35], and systems based on solar spectrum splitting [36]. Enhancement of conversion efficiency was notably modeled by van Sark [28] for multicrystalline silicon PV cells in a retrofitted configuration. Efficiency was shown to critically depend on the heat dissipation at the TE cold side. Koumoto et al. [33] introduced the possibility of enhancing TE efficiency by collecting power through a blackbody interposed between the PV and TE converters (Fig. 7.1). Using a DSSC along with a commercial solar selective absorber (SSA), efficiencies were demonstrated to increase from 9.26% (PV only) to 13.8%.

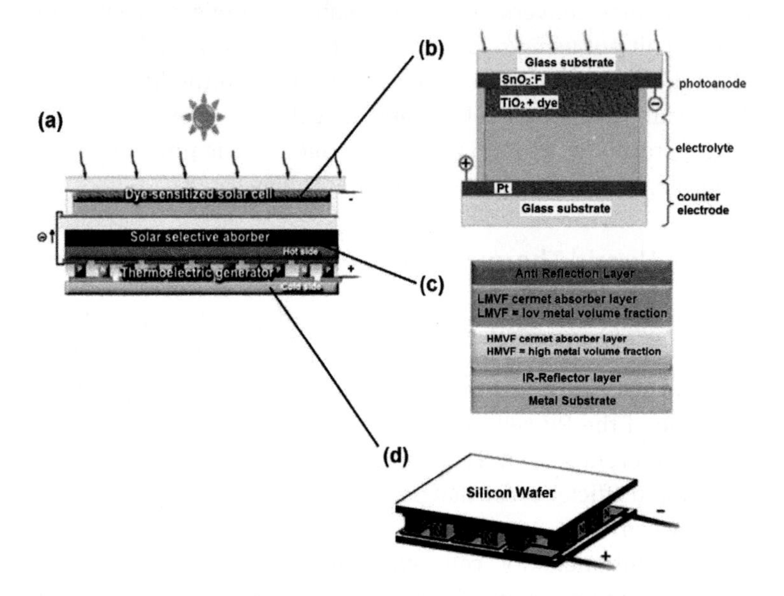


**Figure 7.1** Schematics of a PV-TE hybrid device using a DSSC PV cell coupled to a thermoelectric generator and using a solar selective absorber: (a) the hybrid device, (b) the PV cell, (c) the detailed structure of the solar selective absorber, and (d) the thermoelectric generator. Adapted from Ref. [33] with permission of The Royal Society of Chemistry.

## 7.4 Photovoltaic Generation

A PV cell is essentially a semiconductor diode working under illumination, able to absorb and convert electromagnetic radiation into electrical current. The ability of absorbing the incident radiation depends mainly on the semiconductor characteristics (in particular its absorption coefficient), while the conversion rate is influenced by the device components and the material purity.

In this section we will discuss the physical principles of the interaction between the electromagnetic source (the sun) and the semiconductor material, the main components and properties of the PV device, and the parameters influencing their absorption and conversion efficiencies.

### 7.4.1 Physical Principles

The electromagnetic source of photons relevant to PVs is the sun, which emits a wide photon spectrum. Just above the earth's atmosphere, the incident power is 1366 $W/m^2$ and the spectrum, called air mass zero (AM0), can be approximated by that of a blackbody at 5860 K, as shown in Fig. 7.2. The air mass (AM) index accounts for absorption contribution of the atmosphere on the solar intensity. Figure 7.2 displays the main standard spectrum used for the characterization of PV cells, namely the AM1.5 spectrum with intensity 1000 $W/m^2$.

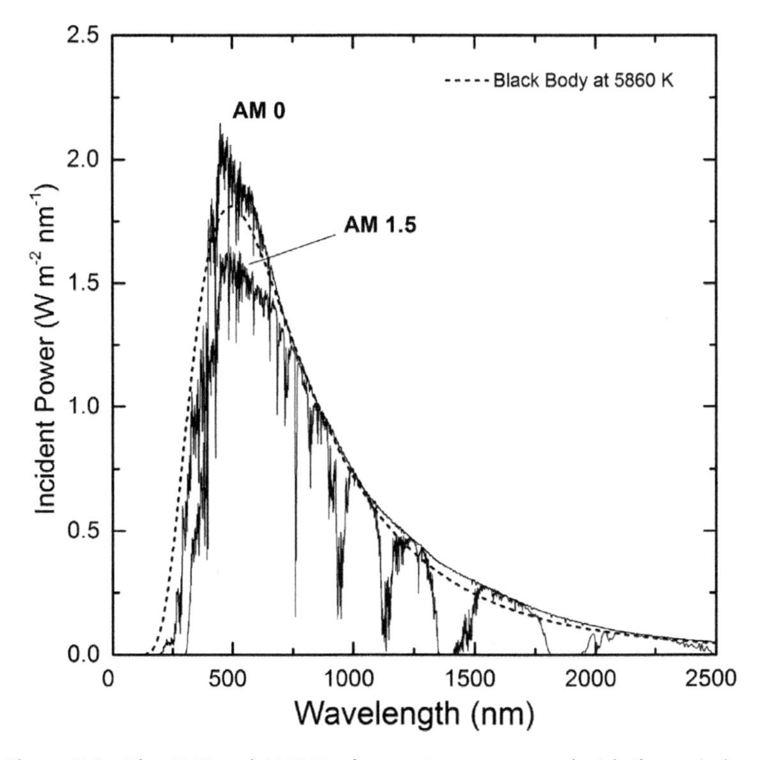


**Figure 7.2** The AM0 and AM1.5 solar spectrum compared with the emission of a blackbody at 5860 K. Data from NREL [37].

The absorption of a photon in semiconductors can cause the excitation of an electron from the material valence band (VB) to the conduction band (CB), leaving an empty state (hole) within the VB. This process, consisting of the creation of an electron-hole pair, has to conserve both the total energy and the momentum of the particles involved. Two different kinds of absorption processes, direct and indirect, are possible in semiconductors. The direct transition is a two-particle process with basically no variation of the electron momentum ($p_{el}$). In fact since the photon momentum $p_\gamma = h/\lambda_\gamma$ of a photon of wavelength $\lambda_\gamma$ (where $h$ is the Planck constant) is several orders of magnitude smaller than $p_{el}$, this transition is nearly perpendicular in the energy/momentum space

$$\begin{aligned} E_{el}^{f} &= E_{el}^{i} + E_\gamma \\ p_{el}^{f} &= p_{el}^{i} + p_\gamma \approx p_{el}^{i} \end{aligned} \tag{7.9}$$

where the superscripts f and i label the final and initial states, respectively.

The indirect transition, instead, is a three-particle process involving also a quantum of the lattice vibration (phonon), which has typically small energy but a large momentum. In this transition the phonon can be either absorbed or emitted and the conservation equations are

$$\begin{aligned} E_{el}^{f} &= E_{el}^{i} + E_\gamma \pm E_{phon} \\ p_{el}^{f} &= p_{el}^{i} + p_\gamma \pm p_{phon} \end{aligned} \tag{7.10}$$

Depending on the semiconductor band structure, and in particular on the mutual position of the VB maximum and the CB minimum in the reciprocal lattice ($\vec{k}$) direction, the energy gap is called direct or indirect. The first case happens when the top of the valence dispersion curve and the bottom of the conduction dispersion curve occur at the same crystal momentum ($\vec{k}$). In materials such as Cu(In,Ga)Se, GaAs, InP, CdTe, CdS, and amorphous silicon ($a$Si), the lowest energy transition is direct. The second case occurs instead in semiconductors such as Si and Ge, where the lowest energy transition is indirect.

Direct photon absorptions are obviously more probable than indirect transitions since they involve a smaller number of particles [38]. Consequently, the absorption coefficient, defined as

$$\alpha \propto \Sigma P^{\mathrm{i,f}} g_{\mathrm{V}}(E^{\mathrm{i}}) g_{\mathrm{C}}(E^{\mathrm{f}}) \quad (7.11)$$

is larger for semiconductors with direct energy gaps, as shown in Fig. 7.3. In Eq. 7.11 $g_V(E^i)$ and $g_C(E^f)$ are respectively the value of the VB density of state at energy $E^i$ and the value of the conduction band density of state at energy $E^f$.

The reverse of the electron-hole pair creation is the so-called recombination process. This event consists in the annihilation of the electron-hole pair, with the exited electron relaxing to its initial state. Recombination may be either radiative or non-radiative. In the radiative case the energy conservation implies the emission of a photon with energy equal to the difference between the excited and relaxed states. Therefore, this process is the inverse of the optical generation process.

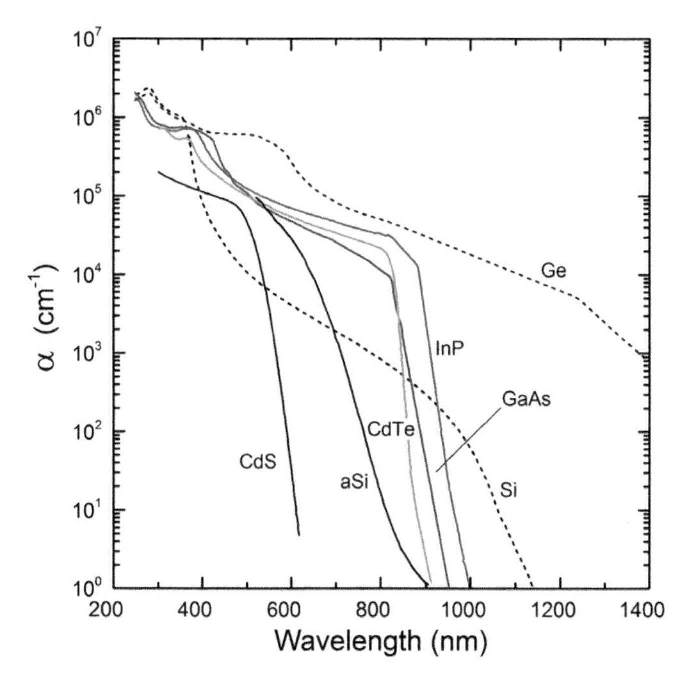


**Figure 7.3** Absorption coefficient versus wavelength for several semiconductors used in photovoltaics.

The rate of such an event can be calculated assuming perfect blackbody absorption above the energy gap and perfect white-body emission below it:

$$R_{\text{rad}} = \frac{2\pi}{c^2 h^3} \int_{E_g}^{\infty} \frac{E^2 \text{d}E}{\exp[E/k_B T] - 1}, \tag{7.12}$$

where $k_B$ and $c$ are the Boltzmann constant and the speed of light, respectively. The material temperature and energy gap are instead $T$ and $E_g$, respectively. Note that Eq. 7.12 shows how radiative recombination is more probable for small energy gaps and for high temperatures.

Nonradiative recombination is instead a trap-assisted process and includes both Shockley–Read–Hall (SHR) and Auger recombination. SHR recombination is due to energy levels within the bandgap, due to impurities and crystal imperfections. Thus the SHR recombination rate depends on the quality of the material and on the position of the impurity levels within the gap. If $N_T$ and $E_T$ are, respectively, the trap density and the trap energy level [39] one finds that

$$R_{\text{SHR}} = \frac{pn - n_i^2}{\dfrac{p + n_i e^{(E_i - E_T)/k_B T}}{\varsigma_n v_{\text{th}} N_T} + \dfrac{n + n_i e^{(E_T - E_i)/k_B T}}{\varsigma_p v_{\text{th}} N_T}}, \tag{7.13}$$

where $p$, $n$, and $n_i^2$ are, respectively, the hole, the electron, and intrinsic carrier densities; $\varsigma_n$ and $\varsigma_p$ are the electron and hole cross sections, respectively; and $v_{\text{th}}$ is the carrier thermal velocity. Combining the cross section with the velocity and the trap density one finds that the carrier lifetimes may be written as

$$\tau_n = \frac{1}{\varsigma_n v_{\text{th}} N_T} \tag{7.14}$$

and

$$\tau_p = \frac{1}{\varsigma_p v_{\text{th}} N_T}. \tag{7.15}$$

Auger recombination is, instead, in some way, similar to the radiative recombination process, save that the energy lost by the carrier is transferred to another carrier, which afterward thermally relaxes, transferring, in turn, its energy to the lattice. The Auger recombination rate includes then the density of either electrons or holes:

$$R_{\text{aug}} = \Lambda_n n(pn - n_i^2) + \Lambda_p p(pn - n_i^2) \tag{7.16}$$

where $\Lambda_n$ and $\Lambda_p$ are the electron and hole Auger coefficients, respectively, for the given semiconductor [39].

Temperature affects the recombination rate not only directly but also through the modulation of the energy gap. Actually, since both recombination and absorption processes are dependent upon $E_g$, any variation of the energy gap with the temperature will modify the material characteristics. Energy gap dependency on $T$ is mostly due to two mechanisms, namely lattice dilation and electron-phonon interactions. The mostly used equation describing this dependency is the semiempirical Varshni's relation:

$$E_g = E_0 - \frac{aT^2}{T+b}, \tag{7.17}$$

where $a$ and $b$ are fitting parameters [40]. Although Eq. 7.17 satisfactorily fits experimental data for materials and temperature ranges of interest in PV applications, alternate and more accurate analyses are available [41, 42].

### 7.4.2 The p-n Junction

When n-type and p-type semiconductors come into contact a p-n junction (diode) is formed and a dual diffusion of so-called majority carriers is established along the junction. The concentration gradient, therefore, induces an electron flow from the n to the p side and a hole flow from the p to the n side. As soon as they cross the junction, electrons and holes become minority carriers and tend to recombine with the majority carriers of the opposite sign found on the other side of the junction. These two processes of diffusion and recombination create in the proximity of the junction two electrically charged regions due to impurity atoms no longer compensated (neutralized) by the free carriers. Thus, an electric field is generated within this layer, usually referred to as a depletion region. Because of the carrier diffusion the field intensity tends to increase, but in the meantime it also hampers further charge diffusion. Thus, the junction comes to equilibrium when the diffusion is totally balanced by the electric field. The process of carrier diffusion and the build-up of the electric field make the p and n sides equalize their Fermi energies throughout the junction (Fig. 7.4). The potential difference resulting from the junction formation, $V_b$, is given by the difference between the Fermi energy of the $p$ and $n$ sides.

The electrostatics of the junction is governed by Poisson's equation, which relates the electric field and the electrostatic potential to the carrier density

$$\frac{d^2\phi}{dx^2} = -\frac{d\varepsilon}{dx} = \frac{q}{\varepsilon}(n_0 - p_0 + N_A - N_D) \tag{7.18}$$

where $\phi$ is the electrostatic potential; $E$ the electric field; $q$ the electron charge; $\varepsilon$ the semiconductor electric permittivity; $p_0$ and $n_0$, respectively, the hole and electron equilibrium concentrations; and $N_A$ and $N_D$, respectively, the ionized acceptor and donor concentrations.

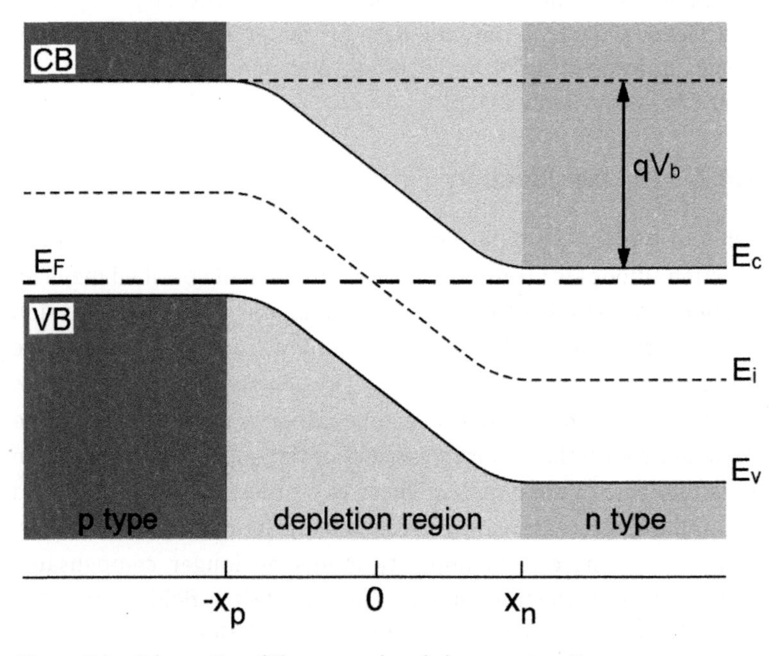


**Figure 7.4** Schematics of the energy bands for a p-n junction.

A standard analysis of Poisson's equation shows that in the depletion layer approximation a potential builds up at the junction [43]:

$$V_b = \frac{qN_A}{2\varepsilon}x_n^2 + \frac{qN_D}{2\varepsilon}x_p^2, \tag{7.19}$$

where $x_n$ and $x_p$ are the depletion widths of the n and p sides of the junction, depending, in turn, on the doping level

$$x_n N_D = x_p N_A, \qquad (7.20)$$

so from Eqs. 7.19 and 7.20 the depletion width depends on the carrier density and $V_b$ as

$$W_D = x_n + x_p = \sqrt{\frac{2\varepsilon}{q}\left(\frac{N_A + N_D}{N_A N_D}\right) V_b}\,. \qquad (7.21)$$

This analysis is valid if there is no external bias applied to the junction. In the case of forward-biased junction (i.e., when the external electric field $V_{ext} > 0$ is opposite to the internal field) the diffusion of majority carriers at the junction is favored, so the depletion region narrows and both the electric field and $V_b$ decrease. In the case instead of a reverse bias ($V_{ext} < 0$) the diffusion of the majority carriers is hindered, so the depletion region widens and both the electric field and $V_b$ increase. Equation 7.21 becomes

$$W_D = x_n + x_p = \sqrt{\frac{2\varepsilon}{q}\left(\frac{N_A + N_D}{N_A N_D}\right)(V_b - V_{ext})} \qquad (7.22)$$

### 7.4.3 Single-Junction Solar Cells: Diode under Illumination

When a diode is under illumination, photons can be absorbed by the processes described in Section 7.4.1. Therefore, if a photon has energy higher than $E_g$ its absorption generates a free electron-hole pair. The junction electric field acts on it by splitting the hole and the electron and making them contributing to the current flowing through the diode. This current enhancement originates from two contributions, namely the electron-hole pairs formed within the depletion layer and the electron-hole pairs formed within a distance smaller than the carrier's diffusion length [43] at both sides of the depletion layer. These two regions are often named quasi-neutral regions. The electron-hole pair generation rate can be written as

$$G(x) = (1 - s)\int_{E_g}^{\infty} (1 - r(E))n_\gamma(E)\alpha(E)e^{-\alpha x}\mathrm{d}E, \qquad (7.23)$$

where $s$ is the grid-shadowing factor, $r(E)$ the reflectivity, $n_\gamma(E)$ the incident photon flux (number of photons per unit of area, time, and energy), and $\alpha(E)$ the absorption coefficient.

The generation rate can be used to compute the light-generated current:

$$I_{\text{gen}} = Aq\int_{x_p - L_p}^{x_n + L_n} G(x)\text{d}x, \tag{7.24}$$

where $A$ is the diode illuminated area and $L_n$ and $L_p$ are the electron and hole diffusion lengths in the pertinent neutral regions. Note that generation occurs both within the depletion layer ($x_p \le x \le x_n$) and in the two quasi-neutral regions ($x_p - L_p \le x \le x_p$ and $x_n \le x \le x_n + L_n$). It is clear from Eqs. 7.23 and 7.24 that $I_{\text{gen}}$ accounts for all the possible optical losses within the solar device.

Using Eq. 7.24 and considering the recombination processes it can be shown that the current flowing within a diode under illumination is

$$I = I_{\text{gen}} - I_{01}(e^{qV/k_BT} - 1) - I_{02}(e^{qV/2k_BT} - 1), \tag{7.25}$$

where $I_{01}$ and $I_{02}$ are, respectively, the dark saturation current due to the recombination in the two quasi-neutral regions and the dark saturation current due to recombination within the depletion region. They are correlated to the recombination processes (cf. Section 7.4.1) as

$$I_{01} = Aq\int_{x_p}^{x_n} R_{\text{tot}}\text{d}x \tag{7.26}$$

and

$$I_{02} = Aq\left[\int_{x_p - L_p}^{x_p} R_{\text{tot}}\text{d}x + \int_{x_n}^{x_n + L_n} R_{\text{tot}}\text{d}x\right], \tag{7.27}$$

where $R_{\text{tot}}$ is the total recombination rate resulting from all active recombination mechanisms.

In view of Eq. 7.25 the solar cell may be modeled as a current source in parallel with two diodes, one with an ideality factor of 1 and the other with an ideality factor of 2. However, often a simpler equivalent circuit is used, encompassing only a single nonideal diode.

The current–voltage characteristics of a typical solar cell are shown in Fig. 7.5. At zero voltage (no external load) the current flowing within the solar cell reads

$$I = I_{\text{SC}}. \tag{7.28}$$

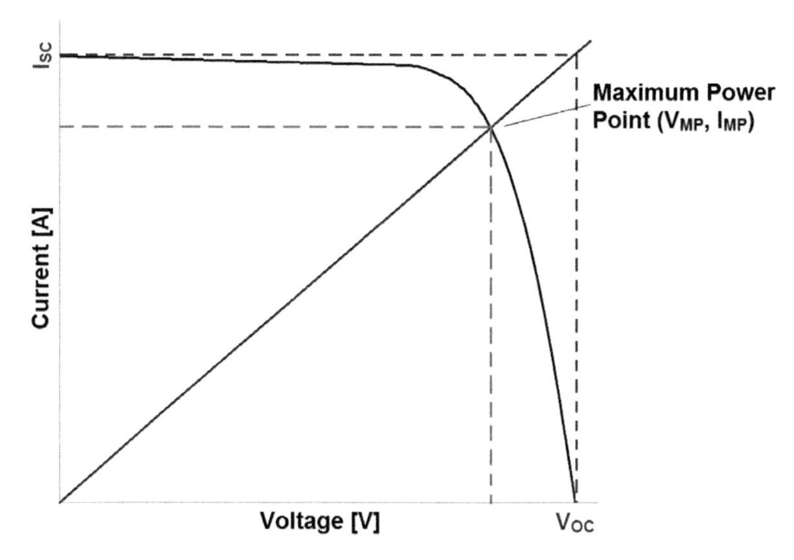


**Figure 7.5** Current–voltage characteristics of a standard solar cell.

This short-circuit current differs from $I_{gen}$ because actually not all the light-generated carriers are collected, as some of them actually recombine before contributing to the cell current. Thus an internal quantum efficiency in collecting light-generated carriers may be defined as the ratio of the short-circuit current and the light-generated current

$$\eta_q^{int} = \frac{I_{SC}}{I_{gen}}. \tag{7.29}$$

The internal quantum efficiency should not be confused with the *external* quantum efficiency, which is defined instead as the ratio between the short-circuit current and the photogenerated current: $\eta_{ext} = I_{SC}$

$$\eta_q^{ext} = \frac{I_{SC}}{I_{ph}}, \tag{7.30}$$

where $I_{ph}$ is the current that would be generated within the cell in the absence of any optical loss. Therefore, $I_{ph} = I_{gen}$ if $r(E) \equiv 0$ and no grid shadowing occurs ($s = 0$).

Under open-circuit conditions (i.e., when no current flows within the device) Eq. 7.25 returns the following expression for the open-circuit voltage $V_{OC}$:

$$V_{OC} = \frac{k_B T}{q} \ln \frac{I_{SC} + I_{01}}{I_{01}} \tag{7.31}$$

The ideal electric power that would be generated by a solar cell having no internal resistance is then $I_{SC}V_{OC}$. Actually, the maximum power of a real solar cell is much smaller due to the internal electric resistance and is given by $I_{MP}V_{MP}$, where the maximum power point (MP) is set so to maximize the $I(V)V$ product (Fig. 7.5). A fill factor (FF) is, therefore, defined as the ratio between the ideal and the actual maximum power outputs:

$$\text{FF} = \frac{I_{MP}V_{MP}}{I_{SC}V_{OC}} \tag{7.32}$$

The parasitic internal resistances in a solar cell are shown in the equivalent circuit of Fig. 7.6. They are a series resistance $R_S$ due mainly to metal contacts (especially from the front grid) and a shunt (parallel) resistance $R_{SH}$ due to alternate electrical pathways for the photogenerated carriers.

When $R_S$ increases the short-circuit current decreases but no effect is produced on the open-circuit voltage. Conversely, when $R_{SH}$ decreases then the open-circuit voltage also decreases, while it does not affect the short-circuit current.

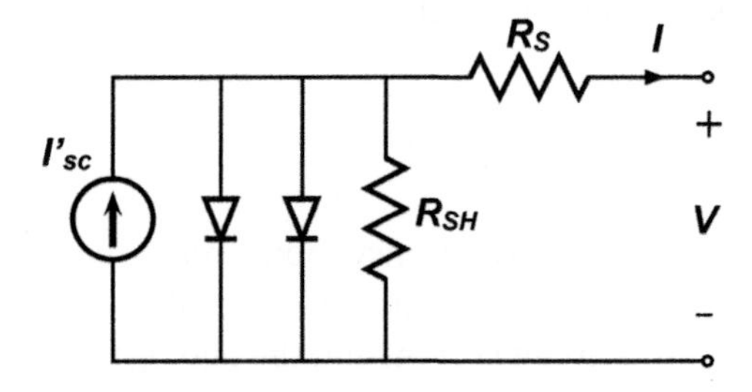


**Figure 7.6** Equivalent electric circuit for a real solar cell with parasitic resistances.

Thus, considering all parasitic resistances and modeling the solar cell in the single-diode approximation Eq. 7.25 becomes

$$I = I_{SC} - I_0[e^{q(V+IR_S)/\beta k_B T} - 1] - \frac{V + IR_S}{R_{SH}}, \tag{7.33}$$

where $\beta$ ranges between 1 and 2, depending on the region where carrier recombination prevalently occurs.

### 7.4.4 Photovoltaic Technology

In what follows the main components of a standard single-junction solar cell are discussed. The most important factors influencing the PV operation will be analyzed, although no attempt will be made to provide a full exhaustive overview of the many kinds of PV cell layouts. More detailed and complete information may be found in standard PV handbooks [44].

The scheme of a general single-junction solar cell is displayed in Fig. 7.7. Moving from the bottom the main components are:

- **Substrate:** It can be rigid or flexible depending on the final use of the solar cell. In general it is made of insulating materials but might also be a metallic foil covered by an insulating film. Its function is to mechanically hold the device. Typical materials used are glass, steel, and polyamides.
- **Back contact:** It is also referred to as rear contact and is the carrier collector. It is engineered to minimize the electrical losses, thus it is supposed to have a small electric resistance. A balance between minimal thickness and low electrical resistance has then to be found. Typically, back contacts are made of silver, aluminum, gold, or molybdenum.

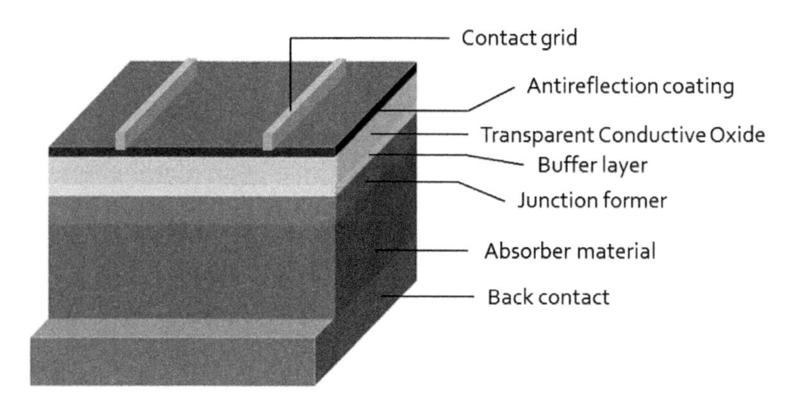


**Figure 7.7** General scheme of a single-junction solar cell with its main components.

- **Absorber material:** Typically of p type, it is the semiconductor within which the photon absorption and the carrier generation occur. The absorber thickness defines the two main categories of single-junction PV devices, namely bulk (thickness of hundreds of micrometers or more) and thin film (thickness of a few micrometers). Typically, bulk materials are indirect-gap absorbers, while thin films use direct-gap semiconductors. The absorber is the material that sets the characteristics required by all other components of the solar cell.

- **Junction former:** Typically of n type, it is the semiconductor that forms along with the absorber material the p-n junction. Its thickness is normally much smaller than that of the absorber material since photons should not be absorbed in this layer.
- **Buffer layer:** Not always present in the PV cells, it is normally used in thin-film solar cells, where it prevents electrical shunts between the absorber material and the upper contacts. It has to be highly transparent, and its characteristics are ruled by defect and doping densities.
- **Transparent front contact:** It is the component that provides the electrical contact between the junction and the contact grid. It should be highly conductive and transparent. Also in this case a trade-off between minimum thickness and lowest electrical resistance has to be found. This layer is normally made of a transparent conductive oxide (TCO) such as $SnO_2$-$In_2O_3$ (ITO) or Al:ZnO (AZO).
- **Antireflective coating:** This component is used to minimize the amount of reflected photons. Usually thin-film interference is used to this aim, growing a $MgF_2$ film on the solar cell top surface. Recently, ZnO nanowire arrays have been shown to be highly effective antireflective coatings [45, 46].
- **Contact grid:** Usually made of a metallic thin film, its standard pattern is characterized by very narrow fingers converging toward a wider strip. The contact area is minimized to prevent excessive light shadowing.

### 7.4.5 PV Efficiency, Energy Gap, and Temperature

From a thermodynamic point of view, the maximum efficiency achievable in a solar cell equals the Carnot efficiency, that is,

$\eta = 1 - T_{cell}/T_{sun}$, where $T_{cell}$ is the solar cell temperature and $T_{sun}$ is the sun temperature. According to Landsberg and Baruch [47], this formula should be modified by taking into account the nonzero radiation emitted by the absorber:

$$\eta_{max} \equiv \frac{P_{max}}{P_{in}} = 1 - \frac{3}{4}\left(\frac{T_{cell}}{T_{sun}}\right) + \frac{1}{3}\left(\frac{T_{cell}}{T_{sun}}\right)^4 \tag{7.34}$$

Then, taking $T_{cell}$ = 300 K and $T_{sun}$ = 6000 K one gets a maximum efficiency of 93.33%. However, Shockley and Queisser (SQ) [48] showed how in a single-junction device the actual achievable efficiency is bound to much lower values due to several types of energy losses. Actually, as already mentioned, while the solar spectrum is continuous over a wide range of frequencies, a PV cell efficiently converts only photons at the frequency corresponding to the energy gap of its absorbing component. Actually, photons with energy lower than $E_g$ are not absorbed while energy of photons with energy larger than $E_g$ is partially converted into heat, thus setting the cell temperature.

The SQ limit neglects optical, thermal, and electrical losses and assumes that $T_{cell}$ equals room temperature. Therefore, $r = 0$, $s = 0$, $R_S = 0$, $1/R_{SH} = 0$, Joule heating is absent, and the device has unitary external quantum efficiency (i.e., any absorbed photon generates an electron-hole pair at a voltage $V_\gamma = E_\gamma/q$). Thus, from Eq. 7.30, $I_{SC} = I_{ph}$. Furthermore, the solar cell is assumed to be oriented normal to the sunlight and the only mechanism of electron-hole recombination is radiative ($R_{aug} = 0$ and $R_{SHR} = 0$).

Following Lorenzi et al. [49, 50] one may evaluate the fractions of energy loss due to photons with $E < E_g$

$$L_{2a}(E_g) = \frac{\int_0^{E_g} n_\gamma(E_\gamma)E_\gamma \mathrm{d}E_\gamma}{\int_0^{\infty} n_\gamma(E_\gamma)E_\gamma \mathrm{d}E_\gamma} \tag{7.35}$$

and that due to photons with $E > E_g$ (causing carrier thermal relaxation)

$$L_{2b}(E_g) = \frac{\int_{E_g}^{\infty} n_\gamma(E_\gamma)(E_\gamma - E_g)\mathrm{d}E_\gamma}{\int_0^{\infty} n_\gamma(E_\gamma)E_\gamma \mathrm{d}E_\gamma}. \tag{7.36}$$

In both equations $n_\gamma$ is the number of photons per unit of area, time, and energy. The total solar power per unit area $\Phi_{sun}$ may be written as

$$\Phi_{sun} = \int_0^\infty n_\gamma(E_\gamma)E_\gamma dE_\gamma. \tag{7.37}$$

Figure 7.8 shows $L_{2a}$ and $L_{2b}$ as a function of the energy gap of the absorber.

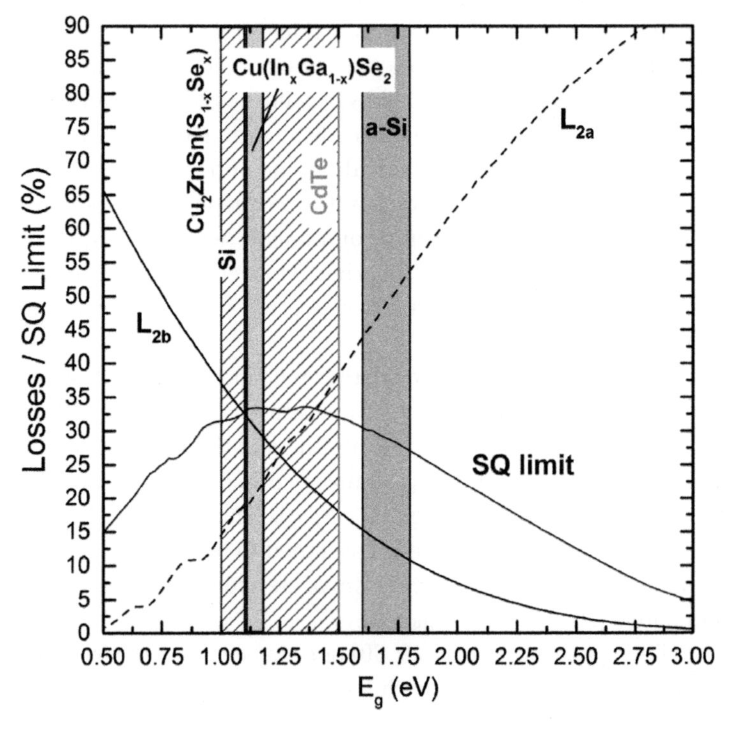


**Figure 7.8** Energy losses $L_{2a}$ and $L_{2b}$ and the SQ limit versus the absorber material energy gap. The solar spectrum was computed using the ASTM data for the Global AM1.5 solar spectrum with an intensity of 1000 W/m$^2$ [37].

Since $I_{SC} = I_{ph}$, using Eq. 7.33 for $R_S = 0$ and $1/R_{SH} = 0$ the current $I$ generated by the solar device considering only radiative recombination is

$$I = I_{ph} - I_0\left[\exp\left(\frac{qV}{\beta k_B T_{cell}}\right) - 1\right], \tag{7.38}$$

where

$$I_{\text{ph}} = Aq\int_{E_{\text{g}}}^{E_{\infty}} n_{\gamma}(E_{\gamma})\text{d}E_{\gamma} \tag{7.39}$$

and since only radiative recombination is supposed to occur, one may write

$$I_0 = AqR_{\text{rad}}. \tag{7.40}$$

Using Eq. 7.38 one then finds the maximum power output $P_{\text{out}}$ by maximizing $I \times V$ for a given $E_{\text{g}}$. The ratio between $P_{\text{out}}(E_{\text{g}})$ and $\Phi_{\text{sun}}$ returns the cell maximum efficiency, namely the SQ limit, as a function of $E_{\text{g}}$ (gray line of Fig. 7.8). The calculation returns that the best PV efficiency in the case of $T_{\text{cell}}$ = 300 K is ~33% for an energy gap between 1.12 and 1.4 eV. This result points out that under these conditions the most viable candidates for PV conversion are silicon ($E_{\text{g}}$ = 1.12 eV), CIGS ($E_{\text{g}}$ = 1.11–1.18 eV), and CZTS ($E_{\text{g}}$ = 1.11–1.4 eV). However, comparing the SQ limit with the experimental solar cell efficiencies (Fig. 7.8), it is evident that the SQ limit highly overestimates the affordable conversion rates. This is mainly due to two assumptions the SQ model relies upon. Firstly, the SQ model assumes an external quantum efficiency equal to 1. This is never the case as reflection from the front contact grid and absorption from the upper finalization layers always occur. The second questionable assumption is that of neglecting nonradiative recombination. In any real material, defects always cause SHR recombination. In addition, Auger recombination will occur, especially in direct-gap semiconductors. Following Green [51], one may extend the SQ model by introducing the so-called external radiative efficiency (ERE), defined as the fraction of the total dark current recombination in the device that results in radiative emission from the device. Thus, the ERE easily accounts for nonradiative recombination by scaling the saturation current $I_0$. As a result Eq. 7.38 may be rewritten as

$$I = I_{\text{ph}} - \frac{I_0}{ERE}\left[\exp\left(\frac{qV}{\beta k_{\text{B}} T_{\text{cell}}}\right) - 1\right]. \tag{7.41}$$

Since the ERE depends on the microstructural quality of a material, it can be actually taken as an index of the technological advancement of a PV material. Figure 7.9 compares the SQ limit, the efficiencies for different values of the ERE, and experimental solar cell efficiencies.

Using Eq. 7.38 for the radiative case and Eq. 7.41 for the nonradiative case one can use the explicit dependence of $I$ and the implicit dependence of $R_{rad}$ (Eq. 7.12) on $T$ to plot the SQ maximum efficiency $\eta(E_g, T)$ as a function of the temperature and of the energy gap, also accounting for the dependence of $E_g$ upon $T$ (Eq. 7.17). This is a relevant correction as changes of $E_g$ modify the integration ranges in Eqs. 7.35, 7.36, and 7.39, and $P_{out}$ and the efficiency therefrom. Since in most semiconductors used in PV devices $E_g$ decreases with the temperature [52], the efficiency dependence on $T$ is the result of two opposite contributions. The photogenerated current $I_{ph}$ increases with $T$ since for smaller $E_g$ more photons generate electron-hole pairs.

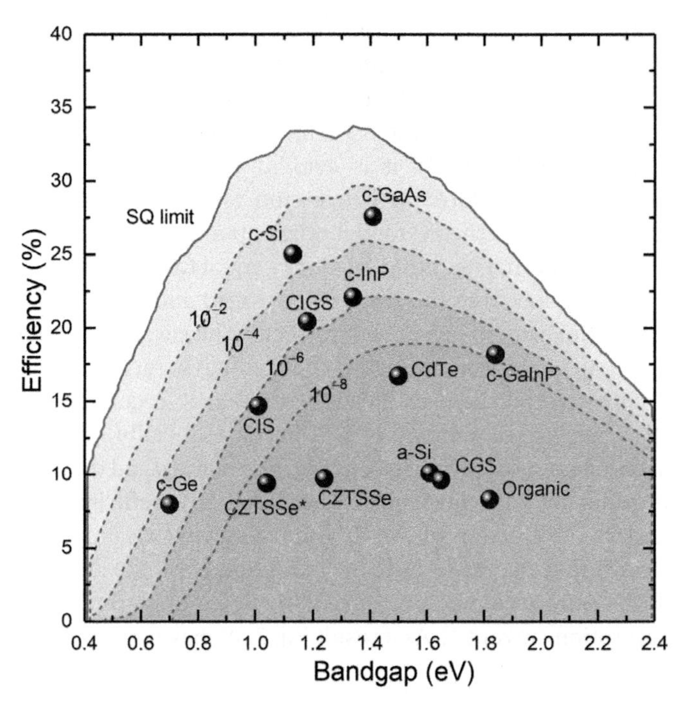


**Figure 7.9** Comparison of the SQ limit and of the efficiency of solar cells for different values of ERE ($1 \times 10^{-2}$, $1 \times 10^{-5}$, and $1 \times 10^{-8}$) with actual PV cell efficiencies.

Conversely, $I_0$ increases with the temperature due to the increasing recombination rate. Therefore, if also the empirical temperature dependence of $E_g$ is taken into account through Eq. 7.17, plotting $\eta$ as

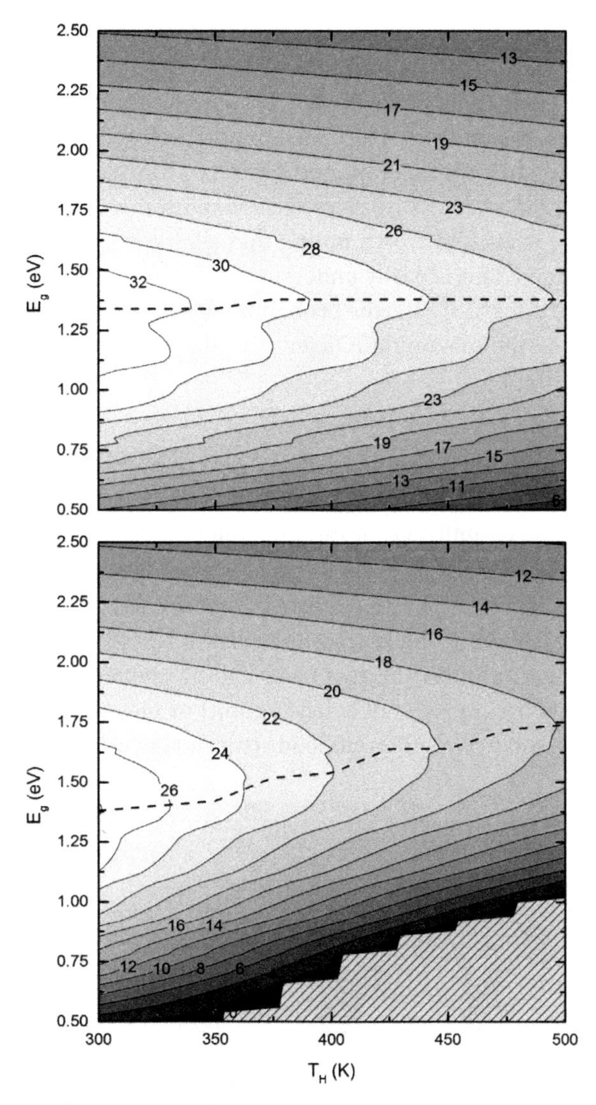


**Figure 7.10** (Top) $\eta(E_g, T_{cell})$ versus $E_g$ and $T_{cell}$ in the radiative limit (ERE = 1); (bottom) the same parameter but with ERE = $1 \times 10^{-3}$, properly fitting c-Si and c-GaAs single-junction solar cells. © [2016] IEEE. Reprinted with permission from Ref. [53].

a function of both temperature and energy gap would be impossible. However, it can be shown that the contribution due to the $E_g$ variation is much smaller compared to that due to the recombination rate.

Thus, neglecting the temperature dependence of the energy gap on $T$ (i.e., assuming a constant value of the energy gap, equal to its value at 300 K) leads to an error of <5% for temperatures up to 500 K in most relevant semiconductors. One may, therefore, compute $\eta_{PV}(E_g, T_{cell})$ under such an assumption (Fig. 7.10). One may immediately conclude that $\eta_{PV}(E_g, T_{cell})$ decreases when temperature increases for any $E_g$. It may be worth noting that such a decrease is steeper for small gaps. This is easily understood considering that, as already mentioned, for small gaps the probability of radiative recombination is higher, so the maximum efficiency value shifts toward higher $E_g$ for increasing $T_{cell}$.

## 7.5 Thermoelectric Hybridization of PV Cells

In this section we will take into account the TE hybridization of single-junction solar cells, based on direct thermal coupling between the PV and TEG parts. The HTEPV layout is reported in Fig. 7.11, where the TEG is shown to be placed underneath the PV device. From a circuital point of view the two parts can be connected in series to a single electrical load (full hybridization) or electrically separated and connected to two different loads (thermal hybridization).

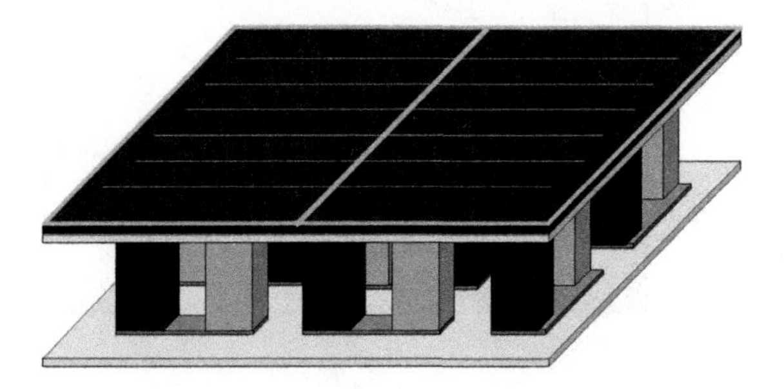

**Figure 7.11** Sketch of the hybrid thermoelectric-photovoltaic device layout.

In the first case, following Park et al. [35], Eq. 7.33 becomes

$$I_{\mathrm{Hyb}} = I_{\mathrm{Hyb}}^{\mathrm{SC}} - I_0 \exp\left\{\frac{q[V - V_{\mathrm{TE}} + I_{\mathrm{Hyb}}(R_{\mathrm{S}} + R_{\mathrm{i,TE}})]}{\beta k_{\mathrm{B}} T} - 1\right\} - \frac{V - V_{\mathrm{TE}} + I_{\mathrm{Hyb}}(R_{\mathrm{S}} + R_{\mathrm{i,TE}})}{R_{\mathrm{Sh}}}, \quad (7.42)$$

where $I_{\mathrm{Hyb}}^{\mathrm{SC}}$ is the HTEPV device short-circuit current and $R_{\mathrm{i,TE}}$ and $V_{\mathrm{TE}}$ are the internal resistance and the TE voltage of the TEG, respectively;

$$R_{\mathrm{i,TE}} = N\left(\frac{\rho_{\mathrm{p}} l}{A_{\mathrm{p}}} + \frac{\rho_{\mathrm{n}} l}{A_{\mathrm{n}}} + R_{\mathrm{c}}\right), \tag{7.43}$$

(where $\rho_{\mathrm{p}}$, $\rho_{\mathrm{n}}$, $A_{\mathrm{p}}$, and $A_{\mathrm{n}}$ are, respectively, the electrical resistivity and the area of the p and n legs; $R_{\mathrm{c}}$ is the contact resistance; $l$ is the length; and $N$ is the number of the TE legs), while

$$V_{\mathrm{TE}} = \int_{T_{\mathrm{C}}}^{T_{\mathrm{H}}} N(\alpha_{\mathrm{p}} - \alpha_{\mathrm{n}})\mathrm{d}T\ , \tag{7.44}$$

where $T_{\mathrm{C}}$ and $T_{\mathrm{H}}$ are the temperatures of the cold and the hot sides of the TEG, while $\alpha_{\mathrm{p}}$ and $\alpha_{\mathrm{n}}$ are, respectively, the Seebeck coefficients of the p and n materials used.

From Eq. 7.42 it can be simply understood how the addition of the TEG in series with the PV cell is expected to lead to electrical losses. In fact the TEG internal electrical resistance $R_{\mathrm{i,TE}}$ enhances the PV cell series resistance $R_{\mathrm{S}}$, thus reducing the short-circuit current. However, Park et al. [35] have recently shown that for any given $R_{\mathrm{i,TE}}$ a minimum $\Delta T = T_{\mathrm{H}} - T_{\mathrm{C}}$ exists for which $V_{\mathrm{TE}}$ is sufficient to compensate the electrical loss. In this case the HTEPV output power equals the sum between the PV and TEG output powers:

$$P_{\mathrm{Hyb}}^{\mathrm{out}} = P_{\mathrm{PV}}^{\mathrm{out}} + P_{\mathrm{TEG}}^{\mathrm{out}}. \tag{7.45}$$

Figure 7.12 shows the result reported by Park et al. [35]. Since it has been demonstrated that under suitable conditions the case of full hybridization can be conveniently reduced to the thermal hybridization case, in what follows only the latter will be considered.

The HTEPV model that will be discussed in this section relies upon the following assumptions:

- The PV cell absorber material acts as a blackbody for the portion of the solar spectrum with energy higher than the absorber material $E_{\mathrm{g}}$, so all photons of this portion of the solar spectrum—$L_{2\mathrm{b}}(E_{\mathrm{g}})$—are absorbed.
- The PV cell absorber material acts as a white body for the portion of the solar spectrum with energy lower than the

absorber material $E_g$, so all photons of this portion of the solar spectrum—$L_{2a}(E_g)$—are not absorbed.

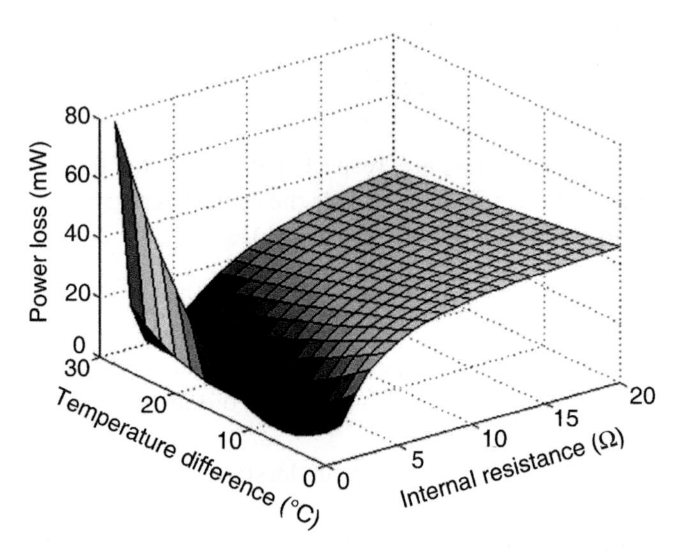


**Figure 7.12** Map of the HTEPV device power losses compared to the sole PV device versus temperature difference and $R_{i,TE}$. Reproduced with permission from Ref. [35].

- $E_g$ is independent of the cell temperature $T$cell.
- All the absorbed photons generate an electron-hole pair.
- The effects of the optical losses and of the carrier recombination are accounted for by an ERE value of $1 \times 10^{-3}$, typical of cells that underwent a high technological advancement (e.g., c-Si and c-GaAs).
- There is a perfect thermal contact between the cell and the TEG hot side, so $T_{cell} = T_H$.
- Ideal heat dissipation occurs at the TEG cold side independently of the heat flowing through it, so $T_C = T_a$ (with the ambient temperature $T_a$ set to 300 K).

In all forthcoming numerical evaluations a TEG figure of merit $ZT$ equal to 1 will be assumed. Under these assumptions the conditions for a beneficial coupling between the PV cell and the TEG will be discussed.

### 7.5.1 Conditions for Enhanced Efficiency in HTEPV Devices

We will follow in this section a discussion recently reported by Lorenzi et al. [49]. Two limiting cases are analyzed. Case 1 will consider a TEG recovering only the thermalization of carriers, namely the thermal effect due to $L_{2b}(E_g)$ only (Eq. 7.36). In this case one can write a simple thermal equation in order to determine the equilibrium temperature of the solar cell (thus the TEG hot temperature):

$$L_{2b}(E_g)\Phi_{sun} = \frac{1}{R_{tot}}(T_{cell} - T_a), \quad (7.46)$$

where $R_{tot}$ is the overall thermal resistance between the device and the ambient. Case 2 considers instead a TEG recovering also the part of the solar spectrum with energy smaller than the absorber energy gap, that is, the $L_{2a}(E_g)$ fraction (Eq. 7.35). Thus Eq. 7.46 becomes

$$[L_{2a}(E_g) + L_{2a}(E_g)]\Phi_{sun} = \frac{1}{R_{tot}}(T_{cell} - T_a) \quad (7.47)$$

It may be worth stressing once again that case 1 and case 2 are meant as limiting situations framing real HTEPV systems. Actually, even if it is true that under-the-gap photons are transmitted through the PV absorber without being absorbed, it is also true that the back contact would at least partially absorb and/or reflect the $L_{2a}(E_g)$ fraction (cf. Fig. 7.7). Therefore, case 1 represents the (ideal) case in which the back contact is totally reflective or transparent for $L_{2a}(E_g)$; instead case 2 depicts the case where under-the-gap photons are totally absorbed by the back contact. Since most of the back contacts used in PV devices are made of metallic films with nonzero reflectance and nonzero absorbance, the real situation will be somewhere in between the two cases.

The TEG converts the heat flowing through its TE elements working between $T_{cell}$ and $T_a$. The output power is

$$P_{TEG}^{out} = \eta_{TEG} L_{2b}(E_g)\Phi_{sun} \quad (7.48)$$

for case 1 and

$$P_{TEG}^{out} = \eta_{TEG}[L_{2a}(E_g) + L_{2b}(E_g)]\Phi_{sun} \quad (7.49)$$

for case 2. In both equations $\eta_{TE}$ is given by Eq. 7.8. Thus the hybrid device efficiency reads

$$\eta_{\mathrm{HTEPV}} = \frac{P_{\mathrm{PV}}^{\mathrm{out}} + P_{\mathrm{TEG}}^{\mathrm{out}}}{\Phi_{\mathrm{sun}}}, \tag{7.50}$$

namely

$$\eta_{\mathrm{HTEPV}} = \eta_{\mathrm{PV}} + \eta_{\mathrm{TE}} L_{2b}(E_g) \tag{7.51}$$

for case 1 and

$$\eta_{\mathrm{HTEPV}} = \eta_{\mathrm{PV}} + \eta_{\mathrm{TE}} \left[L_{2a}(E_g) + L_{2b}(E_g)\right] \tag{7.52}$$

for case 2.

To evaluate the enhancement of the device efficiency with the addition of the TEG to the PV cell it is useful to introduce the normalized HTEPV efficiency defined as

$$\tilde{\eta}_{\mathrm{HTEPV}} = \frac{\eta_{\mathrm{HTEPV}}}{\eta_{\mathrm{PV}}(300\ \mathrm{K})}, \tag{7.53}$$

with $\eta_{\mathrm{PV}}(300\ \mathrm{K})$ being the PV efficiency at 300 K, the best working condition for a PV cell. Figure 7.13 compares $\tilde{\eta}_{\mathrm{HTEPV}}$ to the normalized PV efficiency (dashed gray lines) in both cases.

$$\tilde{\eta}_{\mathrm{PV}} = \frac{\eta_{\mathrm{PV}}}{\eta_{\mathrm{PV}}(300\ \mathrm{K})} \tag{7.54}$$

From this comparison we can immediately conclude that for case 1 the addition of the TEG is only to modestly mitigate the decrease of PV efficiency due to the increase of $T_{\mathrm{cell}}$. This mitigation is more effective for small energy gaps since in this case the decrease of PV efficiency is smaller due to the smaller recombination rates. However, considering the range of temperatures normally reached by solar cells under illumination (310–330 K [54]), such advantage seems unlikely to justify the effort needed to add the TEG stage, as recently discussed by Narducci and Lorenzi [53].

Much more relevant is the impact of the TEG in case 2. In such a scenario while for small $E_g$ the situation is similar to case 1, for $E_g \gtrapprox 1.70\ \mathrm{eV}$ around 300 K, and for $E_g \gtrapprox 1.85\ \mathrm{eV}$ at 500 K the TEG contribution is large enough to overcome the decrease of PV efficiency with temperature. Actually, the level line at 1 (Fig. 7.13) sets the threshold above which the TEG addition positively impacts $\eta_{\mathrm{HTEPV}}$. For such values of $E_g$ one actually finds that the overall efficiency may exceed that of the PV cell at 300 K by more than 20%. This result suggests that in this configuration a TEG stage is well worth being implemented. These findings imply that HTEPV devices have the potential to be coupled with existing wide-bandgap solar cells (e.g., a-Si, CdTe, CGS alloys, and organic) and with PV materials of a lower cost, such as $Cu_2S$ [55] ($E_g \approx 2.65$ eV) and $Cu_2O$ [56] ($E_g \approx$ 2.3 eV), normally not considered for PV conversion because of their low intrinsic performances.

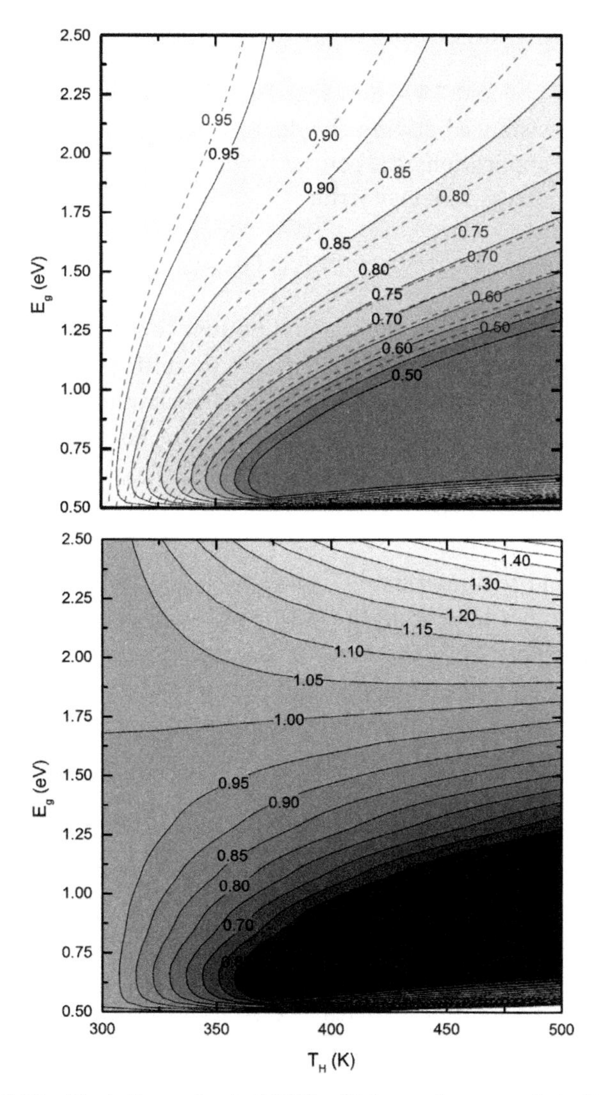


**Figure 7.13** (Top) Normalized HTEPV efficiency for case 1 (solid lines) compared to the normalized PV efficiency (dashed lines); (bottom) normalized HTEPV efficiency for case 2. © [2016] IEEE. Reprinted with permission from Ref. [53].

However, since this enhancement is expected to occur at a range of temperatures around 400 K, actions have to be taken to make the device work at such temperatures. The following sections will discuss the best strategies to satisfy this demand.

### 7.5.2 Optimal Layout of the Thermoelectric Stage

Equations 7.46 and 7.47 show that $T_{cell} = T_H$ is set by $R_{tot}$, the overall thermal resistance between the device and the ambient. Since, for terrestrial applications, the convective heat transfer dominates over radiative process, one may write

$$\frac{1}{R_{tot}} = h_{top} + h_{bot}, \tag{7.55}$$

where $h_{top}$ and $h_{bot}$ are the top and bottom convective coefficients, respectively. As seen, $R_{tot}$ contributes to setting the operational temperature of the HTEPV device. Both a too-low and a too-high $R_{tot}$ lead to the equalization of temperatures at the hot and cold sides of the TEG. For $R_{tot} \to \infty$ one actually gets $T_C \to T_{cell}$, while vanishingly small resistances make $T_{cell} = T_C = T_a$. In both cases the TEG efficiency dramatically decreases.

It should be remarked that in real PV devices heat dissipation actually occurs both from the back and the front surface. Thus, in HTEPV devices both thermal resistances have to be considered. Manifestly enough, efficient hybridization should minimize heat conduction from the PV front, which takes away heat that might be converted through the TEG stage.

Focusing onto case 2 and using Eq. 7.47 one finds that optimal $R_{tot}$ should have values between 0.100 and 0.175 $m^2K/W$ (so that $T_{cell}$ ranges between 375 and 425 K). These values are, however, much higher than those reported for standard PV working conditions. Actually, for freestanding (fs) PV modules the average heat transfer resistance $R^{fs}$ was roughly estimated to be rv 0.021 $m^2K/W$ [57]. Therefore, an additional thermal resistance is needed for $R_{tot}$ to be large enough to enter the range of hybridization profitability. The additional (series) resistance has to be provided by the TEG, namely

$$R_{tot} = R_{PV}^{fs} + R_{TEG}. \tag{7.56}$$

Using Eqs. 7.47 and 7.56 one can plot $R_{\text{TEG}}$ as a function of $E_g$ and $T_{\text{cell}}$.

From Fig. 7.14 one may conclude that $R_{\text{TEG}}$ should range between 0.075 and 0.15 m$^2$K/W. However, standard, commercially available TE devices display thermal resistances between 0.001 and 0.004 m$^2$K/W, namely ~50 times lower than needed. Therefore, one may reach the important conclusion that unless special TEG layouts are introduced, hybridization will fail to achieve its possible efficiency enhancement.

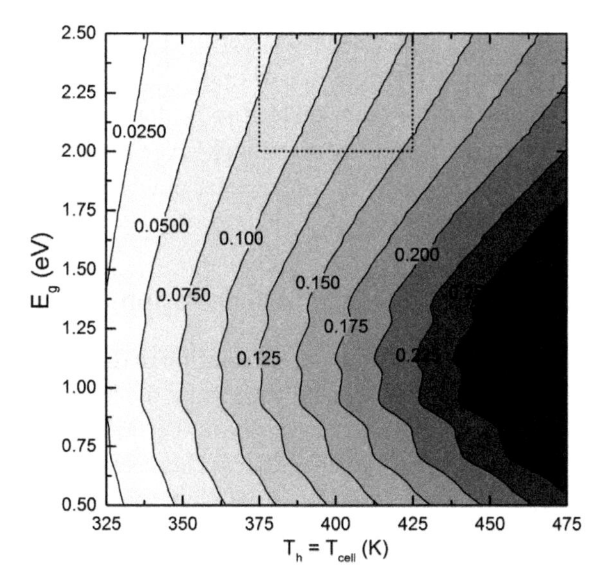


**Figure 7.14** $R_{\text{TEG}}$ (in m$^2$K/W), from Eq. (7.56), versus $E_g$ and $T_{\text{cell}}$. The gray dotted area marks the range of values for which the HTEPV device shows consistently increased performances compared to the pure PV case. © [2016] IEEE. Reprinted with permission from Ref. [53].

Even when properly designed TEGs are available, additional care should be spent to minimize heat dissipation from the front of the PV cell. Manifestly enough, simply increasing the resistivity of the cell back would lead to an increase of heat flowing through the alternate low-resistance path. To prevent it, convective dissipation from the cell front might be avoided by vacuum-sealing it. Thus, dissipation from the front contact would be only radiative, leading to a front thermal resistance of ~0.3 $m^2K/W$ (for $T_{cell}$ between 375 and 425 K and taking $E = 0.3$, a typical value for PV silicon)—namely, a value much larger than that of the (optimized) back thermal resistance. A further improvement may come from heat mirrors (HMs). HMs are optical components with high transmissivity in the UV-Vis. region but high reflectivity in the IR range. As a result they are a proper solution for efficiently transmitting the solar spectrum while ensuring a very small emissivity. On the basis of Fan's results [58] and on the advance of the field of TCOs [59], which are natural candidates for this application, one may safely consider the possibility of properly tuning the cutoff wavelength separating high-transmissivity and high-reflectivity spectral ranges. This can be done by modifying the TCO doping and/or thickness, obtaining cover plates having very little impact (<10%) on the incoming solar power, along with small effective emissivity ($E < 0.01$).

### 7.5.3 Optical and Thermal Concentration

It has been shown that the only viable solution to reach a sufficiently high $T_{cell}$ is to build a TEG able to concentrate the incoming thermal flux. Stated differently, the areal density of the TE legs (often called the filling factor, $FF_{TEG}$) has to be reduced. This is consistent with the conclusions recently reported on solar thermoelectric generators (STEGs) by Kraemer et al. [16]. Defining the thermal concentration

$$C_{th} = \frac{A_{PV}}{A_{TEG}} = \frac{1}{FF_{TEG}}, \tag{7.57}$$

(where $A_{TEG}$ and $A_{PV}$ are, respectively, the TEG active area and the PV cell area) the analysis reported in the previous section shows that $C_{th}$ should be at least 50. Therefore, the TEG FF has to be smaller than $FF_{TEG} = 2\%$, which means (considering no change of the leg heights) that TEGs optimized for HTEPV devices would need less than 4% of the material used in current TEG devices (commercial TEG FF is ~50%).

That would positively lower costs and material utilization, opening interesting prospects for HTEPV devices. However, one should also consider that the reduction of $FF_{TEG}$ implies the increase of the nonactive surface of the TEG. Therefore, the thermal exchange between the hot and the cold TEG surface, not flowing through the TE legs, becomes significant, as pointed out by Kraemer et al. [60] in the case of thermally concentrated STEGs.

As an alternative to thermal concentration, optical concentration could be considered [61]. Actually, such approach has been reported in several studies on STEGs [7, 19, 36, 62]. In this case the increase of $T_{cell}$ is obtained, increasing the input flux in Eq. 7.47. Although this option would enable the use of commercially available TEGs, optical concentration—and more importantly tracking systems—would have to be added to the HTEPV device. This, combined with the use of TEGs with a high $FF_{TEG}$, would surely increase the HTEPV costs, heavily influencing their cost effectiveness. Furthermore, it is well known that the optical concentration increases the PV efficiency. Actually, the short-circuit current becomes

$$I_{SC}^{C_{th}} = C_{th} I_{SC}. \tag{7.58}$$

Thus from Eq. 7.31 and for $I_{SC} \gg I_{01}$ one obtains

$$V_{OC}^{C_{th}} \approx V_{OC} + \frac{k_B T}{q} \ln C_{th} \tag{7.59}$$

so that [44]

$$\eta_{PV}^{C_{th}} \approx \eta_{PV} \frac{FF_{PV}^{C_{th}}}{FF_{PV}} \left( 1 + \frac{k_B T}{q V_{OC}} \ln C_{th} \right). \tag{7.60}$$

Figure 7.15 shows $\eta_{PV}^{C_{th}}$ normalized to the PV efficiency without optical concentration versus Cth, showing how for $C_{opt} > 50$ $\eta_{PV}^{C_{th}}$ is 15% larger than $\eta_{PV}$. Therefore, optical concentration significantly reduces the convenience of adding a TEG stage to the PV cell. This conclusion may be unexpected. However, it simply results from the fact that while $\eta_{TE}$ remains constant, $\eta_{PV}$ increases so the TEG contribution to the overall HTEPV efficiency (Eq. 7.52) becomes smaller. This further reduces the appeal of TEG addition to the PV cell, leading to the conclusion that optical concentration is not implementable or at least not cost effective.

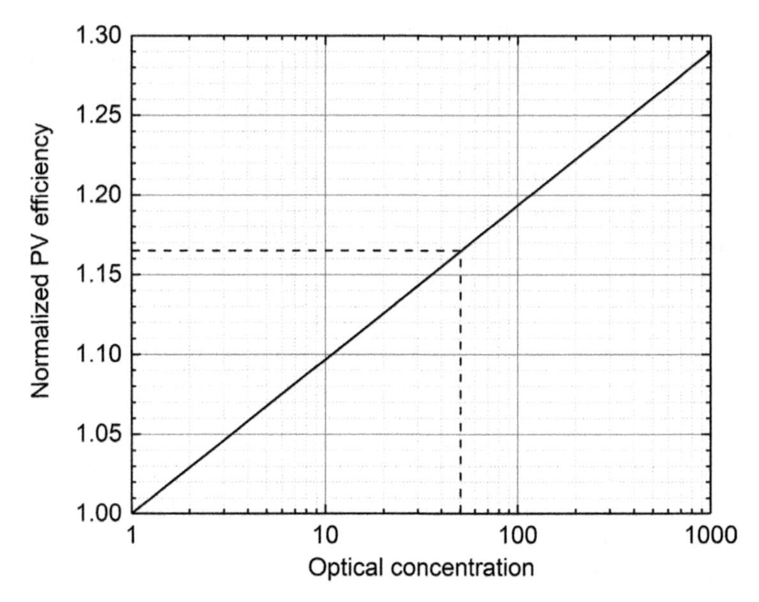


**Figure 7.15** PV efficiency in the case of optical concentration normalized on the PV efficiency without concentration as a function of $C_{th}$.

## 7.6 Summary and Concluding Remarks

In this chapter we have presented the state of the art of HTEPVGs. We have shown that coupling a PV cell with a TEG may result in a sensitive increase of the overall efficiency, especially when large-gap PV materials are used. As a result, HTEPVGs, in addition to enhancing the profitability of standard PV cells, may also promote the use of low-cost PV absorbers not currently employed because of their limited PV efficiency. We have also shown that several issues have to be carefully considered and optimized to achieve effective PV-TE coupling. Specifically, a proper, nonstandard geometry of the TEG has to be developed so as to increase the thermal resistance of the device. This also implies the implementation of thermal concentration technologies.

Compared to STEGs and to hybrid cogenerative solar thermoelectric generators (HCG-STEGs), it is possibly fair to conclude that HTEPVGs are a sensible compromise between closeness to market and technological ambition. HCG-STEGs have already been deployed and are in use in a number of countries. Still, they are basically capable of very modest electrical conversion efficiencies as they retain the architecture of thermal solar converters, implementing TEGs as add-ons. On the opposite side, STEGs pave the way to a visionary technology, capable in principle not only of large electric conversion efficiency (estimated to reach 14% upon full optimization [16]) but also of remarkable material savings. Nonetheless, they critically depend upon the availability of TE materials with $ZT \geq 2$ to fully compete with current PV technologies. HTEPVGs stand in between, taking advantage of the high conversion efficiencies achieved by a host of materials, along with the PV technological maturity, still enhancing their power outputs by using the heat PV cells dissipate. Thus, though HTEPVGs also have significant technological hurdles to be overcome, they are possibly closer than STEGs to hit the solar market. Compared instead to advanced, high-efficiency PV cells using optical concentration, multijunctions, and solar spectrum engineering (up and down conversion), HTEPVGs might provide a more convenient route to scale up conversion efficiency without scaling up manufacturing complexity as well.

## Acknowledgments

This work was supported by FP7–NMP–2013–SMALL-7, SiN-ERGY (Silicon Friendly Materials and Device Solutions for Microenergy Applications), Contract No. 604169.

## References

1. NREL. (2015). NREL photovoltaics efficiency chart, http://www.nrel.gov/ncpv/images/efficiency_chart.jpg, accessed: 2015-08-20.
2. EnergyTrend. (2015). http://pv.energytrend.com/.
3. Repins, I., Contreras, M., Egaas, B., DeHart, C., Scharf, J., Perkins, C., To, B., and Noufi, R. (2008). 19.9%-efficient zno/cds/cuingase2 solar cell with 81.2% fill factor, *Prog. Photovoltaics Res. Appl.*, **16**(3), pp. 235–239, doi:10.1002/pip.822, http://www.scopus.com/inward/record.url?eid=2-s2.0-42249114488&partnerID=40&md5=33f37c95c84cf6f175496f222433956c.

4. Mitzi, D., Gunawan, O., Todorov, T., Wang, K., and Guha, S. (2011). The path towards a high-performance solution-processed kesterite solar cell, *Sol. Energy Mater. Sol. Cells*, **95**(6), pp. 1421–1436, doi:10.1016/j.solmat.2010.11.028, http://www.scopus.com/ inward/record.url?eid=2-s2.0-79955007801&partnerID=40&md5=c3561287cd559425128bab412dbe1925.

5. De Groot, S., and Mazur, P. (2013). *Non-Equilibrium Thermodynamics*, Dover, https://books.google.it/books?id=mfFyG9jfaMYC.

6. Ioffe, A. (1957). *Semiconductor Thermoelements and Thermoelectric Cooling*, Infosearch, London, UK.

7. Telkes, M. (1954). Solar thermoelectric generators, *J. Appl. Phys.*, **25**(6), pp. 765–777, doi:10.1063/1.1721728.

8. Chapin, D. M., Fuller, C. S., and Pearson, G. L. (1954). A new silicon p–n junction photocell for converting solar radiation into electrical power, *J. Appl. Phys.*, **25**(5), pp. 676–677, doi:http://dx.doi.org/10.1063/1.1721711, http://scitation.aip.org/content/aip/journal/jap/25/5/10.1063/1.1721711.

9. Kanatzidis, M. G. (2009). Nanostrúctured thermoelectrics: the new paradigm? *Chem. Mater.*, **22**(3), pp. 648–659.

10. Heremans, J. P., Dresselhaus, M. S., Bell, L. E., and Morelli, D. T. (2013). When thermoelectrics reached the nanoscale, *Nat. Nanotech.*, **8**(7), pp. 471–473.

11. Zhao, L.-D., Lo, S.-H., Zhang, Y., Sun, H., Tan, G., Uher, C., Wolverton, C., Dravid, V. P., and Kanatzidis, M. G. (2014). Ultralow thermal conductivity and high thermoelectric figure of merit in SnSe crystals, *Nature*, **508**(7496), pp. 373–377.

12. Hochbaum, A. I., Chen, R. K., Delgado, R. D., Liang, W. J., Garnett, E. C., Najarian, M., Majumdar, A., and Yang, P. D. (2008). Enhanced thermoelectric performance of rough silicon nanowires, *Nature*, **451**(7175), pp. 163–167.

13. Boukai, A. I., Bunimovich, Y., Tahir-Kheli, J., Yu, J. K., Goddard, W. A., and Heath, J. R. (2008). Silicon nanowires as efficient thermoelectric materials, *Nature*, **451**(7175), pp. 168–171.

14. Narducci, D., Frabboni, S., and Zianni, X. (2015). Silicon de novo: energy filtering and enhanced thermoelectric performances of nanocrystalline silicon and silicon alloys, *J. Mater. Chem. C*, **3**, pp. 12176–12185, doi:10.1039/C5TC01632K, http://dx. doi.org/10.1039/C5TC01632K.

15. Zhao, L.-D., Tan, G., Hao, S., He, J., Pei, Y., Chi, H., Wang, H., Gong, S., Xu, H., Dravid, V. P., Uher, C., Snyder, G. J., Wolverton, C., and Kanatzidis,

M. G. (2015). Ultrahigh power factor and thermoelectric performance in hole-doped single-crystal SnSe, *Science*, doi:10.1126/science.aad3749, http://www.sciencemag.org/content/early/2015/11/24/science.aad3749.abstract.

16. Kraemer, D., Poudel, B., Feng, H.-P., Caylor, J. C., Yu, B., Yan, X., Ma, Y., Wang, X., Wang, D., Muto, A., McEnaney, K., Chiesa, M., Ren, Z., and Chen, G. (2011). High-performance flat-panel solar thermoelectric generators with high thermal concentration, *Nat. Mater.*, **10**(7), pp. 532–538, doi:10.1038/nmat3013, http://dx.doi.org/10.1038/nmat3013.
17. Goldsmid, H., Giutronich, J., and Kaila, M., (1980). Solar thermoelectric generation using bismuth telluride alloys, *Sol. Energy*, **24**, p. 435440.
18. Omer, S. (1998). Design optimization of thermoelectric devices for solar power generation, *Sol. Energy Mater. Sol. Cells* **53**, 1-2, pp. 67–82, doi:10.1016/S0927-0248(98)00008-7.
19. Amatya, R., and Ram, R. J. (2010). Solar thermoelectric generator for micropower applications, *J. Electron. Mater.*, **39**(9), pp. 1735–1740, doi:10.1007/s11664-010-1190-8.
20. Mizoshiri, M., Mikami, M., Ozaki, K., and Kobayashi, K. (2012b). Thin-film thermoelectric modules for power generation using focused solar light, *J. Electron. Mater.*, **41**(6), pp. 1713–1719, doi:10.1007/s11664-012-2047-0.
21. Suter, C., Tomeš, P., Weidenkaff, A., and Steinfeld, A. (2011). A solar cavity-receiver packed with an array of thermoelectric converter modules, *Sol. Energy*, **85**(7), pp. 1511–1518, doi:10.1016/j.solener.2011.04.008.
22. Rockendorf, G., Sillmann, R., Podlowski, L., and Litzenburger, B. (1999). PV-hybrid and thermoelectric collectors, *Sol. Energy*, **67**(4–6), pp. 227–237, doi:10.1016/S0038-092X(00)00075-X.
23. Fan, H., Singh, R., and Akbarzadeh, A. (2011). Electric power generation from thermoelectric cells using a solar dish concentrator, *J. Electron. Mater.*, **40**(5), pp. 1311–1320, doi:10.1007/s11664-011-1625-x.
24. Miljkovic, N., and Wang, E. N. (2011). Modeling and optimization of hybrid solar thermoelectric systems with thermosyphons, *Sol. Energy*, **85**(11), pp. 2843–2855, doi:10.1016/j.solener.2011.08.021, http://dx.doi.org/10.1016/j.solener.2011.08.021.
25. Chávez Urbiola, E. A., and Vorobiev, Y. (2013). Investigation of solar hybrid electric/thermal system with radiation concentrator and thermoelectric generator, *Int. J. Photoenergy*, **2013**(4), doi:10.1155/2013/704087.

26. Zhang, M., Miao, L., Kang, Y. P., Tanemura, S., Fisher, C. a. J., Xu, G., Li, C. X., and Fan, G. Z. (2013a). Efficient, low-cost solar thermoelectric cogenerators comprising evacuated tubular solar collectors and thermoelectric modules, *Appl. Energy*, **109**, pp. 51–59, doi:10.1016/j.apenergy.2013.03.008, http://dx.doi.org/10.1016/j.apenergy.2013.03.008.

27. Vorobiev, Y., González-Hernández, J., Vorobiev, P., and Bulat, L. (2006). Thermal-photovoltaic solar hybrid system for efficient solar energy conversion, *Sol. Energy*, **80**, pp. 170–176, doi:10.1016/j.solener.2005.04.022.

28. Van Sark, W. G. J. H. M. (2011). Feasibility of photovoltaic–thermoelectric hybrid modules, *Appl. Energy*, **88**, pp. 2785–2790, doi:10.1016/j.apenergy.2011.02.008.

29. Li, Y., Witharana, S., Cao, H., Lasfargues, M., Huang, Y., and Ding, Y. (2014). Wide spectrum solar energy harvesting through an integrated photovoltaic and thermoelectric system, *Particuology*, **15**, pp. 39–44, doi:10.1016/j.partic.2013.08.003, http://dx.doi.org/10.1016/j.partic.2013.08.003.

30. Fisac, M., Villasevil, F. X., and López, A. M. (2014). High-efficiency photovoltaic technology including thermoelectric generation, *J. Power Sources*, **252**, pp. 264–269, doi:10.1016/j.jpowsour.2013.11.121, http://dx.doi.org/10.1016/j.jpowsour.2013.11.121.

31. Liao, T., Lin, B., and Yang, Z. (2014). Performance characteristics of a low concentrated photovoltaic thermoelectric hybrid power generation device, *Int. J. Therm. Sci.*, **77**, pp. 158–164, doi:10.1016/j.ijthermalsci.2013.10.013, http://www.sciencedirect.com/science/article/pii/S1290072913002494.

32. Guo, X. Z., Zhang, Y. D., Qin, D., Luo, Y. H., Li, D. M., Pang, Y. T., and Meng, Q. B. (2010). Hybrid tandem solar cell for concurrently converting light and heat energy with utilization of full solar spectrum, *J. Power Sources*, **195**(22), pp. 7684–7690, doi:10.1016/j.jpowsour.2010.05.033, http://dx.doi.org/10.1016/j.jpowsour.2010.05.033.

33. Wang, N., Han, L., He, H., Park, N.-H., and Koumoto, K. (2011). A novel high-performance photovoltaic-thermoelectric hybrid device, *Energy Environ. Sci.*, **4**, p. 3676, doi:10.1039/c1ee01646f.

34. Zhang, Y., Fang, J., He, C., Yan, H., Wei, Z., and Li, Y. (2013b). Integrated energy-harvesting system by combining the advantages of polymer solar cells and thermoelectric devices, *J. Phys. Chem. C*, **117**(47), pp. 24685–24691, doi:10.1021/jp4044573.

35. Park, K.-T., Shin, S.-M., Tazebay, A. S., Um, H.-D., Jung, J.-Y., Jee, S.-W., Oh, M.-W., Park, S.-D., Yoo, B., Yu, C., and Lee, J.-H. (2013). Lossless hybridization between photovoltaic and thermoelectric devices, *Sci. Rep.*, **3**, pp. 422–427, doi:10.1038/srep02123, http://www.pubmedcentral.nih.gov/articlerender.fcgi?artid=3699810{\&}tool=pmcentrez{\&}rendertype=abstract.

36. Mizoshiri, M., Mikami, M., and Ozaki, K. (2012a). Thermal-photovoltaic hybrid solar generator using thin-film thermoelectric modules, *Jpn. J. Appl. Phys., Part 2*, **51**(6), doi:10.1143/JJAP.51.06FL07.

37. ASTM. (2012). ASTM G173-03(2012), Standard Tables for Reference Solar Spectral Irradiances: Direct Normal and Hemispherical on 37 Tilted Surface, http://rredc.nrel.gov/solar/spectra/am1.5/.

38. Pankove, J. I. (1971). *Optical Processes in Semiconductors*, Courier, https://books.google.com/books?id=HHM9Vo0DYZAC{\&}pgis=1.

39. Sze, S. M. (1981). *Physics of Semiconductor Devices*, 2nd edn., Wiley, New York, NY.

40. Varshni, Y. (1967). Temperature dependence of the energy gap in semiconductors, *Physica*, **34**(1), pp. 149–154, doi:10.1016/0031-8914(67)90062-6.

41. O'Donnell, K. P., and Chen, X. (1991). Temperature dependence of semiconductor band gaps, *Appl. Phys. Lett.*, **58**(25), pp. 2924–2926, doi:10.1063/1.104723.

42. Ünlu¨, H. (1992). A thermodynamic model for determining pressure and temperature effects on the bandgap energies and other properties of some semiconductors, *Solid-State Electron.*, **35**(9), pp. 1343–1352, doi:10.1016/0038-1101(92)90170-H, http://www.sciencedirect.com/science/article/pii/003811019290170H.

43. Dalven, R. (1980). *Introduction to Applied Solid State Physics: Topics in the Applications of Semiconductors, Superconductors, and the Nonlinear Optical Properties of Solids*, 1st edn., Springer Science & Business Media, https://books. google.com/books?id=rN7gBwAAQBAJ{\&}pgis=1.

44. Gray, J. L. (2003). The physics of the solar cell, in *Handbook of Photovoltaic Science and Engineering*, pp. 61–111, edited by L. Antonio and H. Steven, John Wiley and Sons.

45. Aurang, P., Demircioglu, O., Es, F., Turan, R., and Unalan, H. E. (2013). ZnO nanorods as antireflective coatings for industrial-scale single-crystalline silicon solar cells, *J. Am. Ceram. Soc.*, **96**(4), pp. 1253–1257, doi:10.1111/jace.12200, http://doi.wiley.com/10.1111/jace.12200.

46. Lee, J. W., Ye, B. U., Kim, D.-Y., Kim, J. K., Heo, J., Jeong, H. Y., Kim, M. H., Choi, W. J., and Baik, J. M. (2014). ZnO nanowire- based antireflective coatings with double-nanotextured surfaces, *ACS Appl. Mater. Interfaces*, **6**(3), pp. 1375–1379, doi:10.1021/am4051734, http://dx.doi.org/10.1021/am4051734.

47. Landsberg, P. T., and Baruch, P. (1989). The thermodynamics of the conversion of radiation energy for photovoltaics, *J. Phys. A*, **22**, 11, p. 1911, http://stacks.iop.org/0305-4470/22/i=11/ a=028.

48. Shockley, W., and Queisser, H. J. (1961). Detailed balance limit of efficiency of p-n junction solar cells, *J. Appl. Phys.*, **32**(1961), pp. 422–427, doi:10.1063/1.1736034.

49. Lorenzi, B., Acciarri, M., and Narducci, D. (2015a). Analysis of thermal losses for a variety of single-junction photovoltaic cells: an interesting means of thermoelectric heat recovery, *J. Electron. Mater.*, **44**(6), pp. 1809–1813, doi:10.1007/s11664-014-3562-y.

50. Lorenzi, B., Acciarri, M., and Narducci, D. (2015b). Conditions for beneficial coupling of thermoelectric and photovoltaic devices, *J. Mater. Sci.*, **30**(17), pp. 2663–2669, doi:10.1557/jmr.2015.174.

51. Green, M. A. (2012). Radiative efficiency of state-of-the-art photovoltaic cells, *Prog. Photovoltaics Res. Appl.*, **20**(4), pp. 472–476, doi:10.1002/pip.1147, http://doi.wiley.com/10.1002/pip.1147.

52. Dupr´e, O., Vaillon, R., and Green, M. (2015). Physics of the temperature coefficients of solar cells, *Sol. Energy Mater. Sol. Cells*, **140**, pp. 92–100, doi:http://dx.doi.org/10.1016/j.solmat.2015.03.025, http://www.sciencedirect.com/science/article/pii/S0927024815001403.

53. Narducci, D., and Lorenzi, B. (2016). Challenges and perspectives in tandem thermoelectric-photovoltaic solar energy conversion, *IEEE Trans. Nanotechnol.*, **15**, pp. 348–355.

54. Nordmann, T., and Clavadetscher, L. (2003). Understanding temperature effects on PV system performance, in *Proceedings of the 3rd World Conference on Photovoltaic Energy Conversion*, Vol. 3, pp. 2–5.

55. Saraf, R. (2012). High Efficiency and cost effective Cu 2 S / CdS thin-film solar cell, *J. Electron. Eng.*, **2**(4), pp. 47–51.

56. Kumar, V., Masudy-Panah, S., Tan, C., Wong, T., Chi, D., and Dalapati, G. (2013). Copper oxide based low cost thin film solar cells, in *2013 IEEE 5th International Nanoelectronics Conference (INEC)*, pp. 443–445, doi:10.1109/INEC.2013.6466072.

57. Skoplaki, E., and Palyvos, J. (2009). Operating temperature of photovoltaic modules: a survey of pertinent correlations, *Renew. Energy,* **34**(1), pp. 23–29, doi:http://dx.doi.org/10.1016/j.renene.2008.04.009, http://www.sciencedirect.com/science/article/pii/S0960148108001353.

58. Fan, J. C., and Bachner, F. J. (1976). Transparent heat mirrors for solar-energy applications, *Appl. Opt.,* **15**(4), pp. 1012–1017, doi:10.1364/AO.15.001012.

59. Ginley, D., Hosono, H., and Paine, D. C. (2010). *Handbook of Transparent Conductors,* Springer Science & Business Media, https://books.google.com/books?id= K0qjBlrAGYsC{\&}pgis=1.

60. Kraemer, D., McEnaney, K., Chiesa, M., and Chen, G. (2012). Modeling and optimization of solar thermoelectric generators for terrestrial applications, *Sol. Energy,* **86**(5), pp. 1338–1350, doi:10.1016/j.solener.2012.01.025, http://dx.doi.org/10.1016/j.solener.2012.01.025.

61. Weinstein, L. A., Loomis, J., Bhatia, B., Bierman, D. M., Wang, E. N., and Chen, G. (2015). Concentrating solar power, *Chem. Rev.,* **115**(23), pp. 12797–12838.

62. McEnaney, K., Kraemer, D., Ren, Z., and Chen, G. (2011). Modeling of concentrating solar thermoelectric generators, *J. Appl. Phys.,* **110**(7), doi:10.1063/1.3642988.

# Chapter 8

# Phonon Transport Effects in Ultranarrow, Edge-Roughened Graphene Nanoribbons

**Neophytos Neophytou[a] and Hossein Karamitaheri[b]**
[a]*School of Engineering, University of Warwick, Coventry CV4 7AL, UK*
[b]*Department of Electrical Engineering, University of Kashan, Kashan 87317-53153, Iran*
n.neophytou@warwick.ac.uk

In this chapter we investigate the influence of low dimensionality and disorder in phonon transport in ultranarrow armchair graphene nanoribbons (AGNRs) using nonequilibrium Green's function (NEGF) simulation techniques. We specifically focus on how different parts of the phonon spectrum are influenced by geometrical confinement and line edge roughness. The discussion and findings should be perceived as general features of phonon transport in purely 1D disordered systems. We discuss how the introduction of line edge roughness does not significantly affect the acoustic modes, which still remain the main heat carriers, but how it significantly affects regions of the phonon spectrum with a low density of phonon modes. In that case, roughness can introduce a band mismatch or, in the worst case, "effective transport gaps," which greatly increase with

*Nanophononics: Thermal Generation, Transport, and Conversion at the Nanoscale*
Edited by Zlatan Aksamija

ISBN 978-981-4774-41-3 (Hardcover), 978-1-315-10822-3 (eBook)
www.panstanford.com

channel length and force phonons into the localization transport regime.

## 8.1 Introduction

The thermal properties of graphene nanostructures and low-dimensional channels in general is an important topic of nanoscience. Graphene nanoribbons (GNRs) are 1D structures that have attracted significant attention, both for fundamental research as well as for technological applications [1–14]. Ultranarrow GNRs have been shown to retain to some degree the remarkable thermal properties of graphene. However, the width, the chirality, the presence of edges, and the magnitude of edge disorder can result in geometry-dependent properties. These parameters can strongly determine a GNR's heat transport properties [9, 10, 15–17], in addition to its electronic properties [18–21].

Several works have shown that the transport properties of low-dimensional systems are significantly degraded by the introduction of scattering centers and localized states [9, 10, 14, 22–25]. In the case of electronic transport, even a small degree of disorder can drastically reduce the electronic conductivity (especially in armchair graphene nanoribbons [AGNRs] rather than zigzag graphene nanoribbons [ZGNRs]), even driving carriers into the localization regime, and introduce effective transmission bandgaps [21, 26–28]. Line edge roughness (LER) can have a similar effect on the thermal properties of GNRs [29], and in this chapter we theoretically explore this in depth. Carbon-related materials such as graphene, nanotubes, and GNRs can have huge thermal conductivities in their pristine form, reaching values as high as of 3080–5150 W/m-K at room temperature [2, 30]. Even a small degree of disorder, however, can drastically degrade this superior thermal conductivity.

Recent theoretical studies attempt to address the thermal properties of low-dimensional materials by employing a variety of models and techniques depending on the size of the channel and the physical effects under consideration. Methods to investigate low-dimensional thermal transport vary from molecular dynamics [25, 31–35], the Boltzmann transport equation (BTE) for phonons using scattering rates based on the single-mode relaxation time approximation (SMRTA) [36–42], the nonequilibrium Green's function (NEGF) method [14, 16, 24, 43–47], and the Landauer method [48–51], but also even more simplified semianalytical methods that employ the Casimir formula to extract boundary scattering rates by assigning a diffusive or specular nature to the boundaries [52, 53].

Tremendous theoretical and experimental investigations of thermal conductivity in nanostructures can be found in the literature. The reason why the phonon transport properties of low-dimensional channels in general and carbon-based systems in particular are recently receiving much attention is the fact that they show certain features that are distinct from bulk materials. From a computational point of view, thermal transport studies in graphene and GNRs, specifically, are in general more tangible compared to other low-dimensional systems (e.g., nanowires) because they involve a smaller number of atoms in the computational domain and can still illuminate the physics of phonon transport at the nanoscale. Several experimental and theoretical works suggest that thermal conductivity could deviate from Fourier's law [3, 12, 54]. It was observed that it grows monotonically with channel length before it saturates at large channel lengths, even lengths significantly larger than the average mean free path (MFP) [8, 55], an indication of a crossover from ballistic into diffusive transport regimes [56, 57]. A recent theoretical study showed that, in the case of pristine 1D channels, the thermal conductivity could even increase with confinement [58]. References [59–61] demonstrate that the thermal conductivity in 1D channels grows as a power-law function of the length and that roughness affects the value of the exponent of this dependence. In 2D graphene channels, on the other hand, the increase in thermal conductivity with channel length follows a logarithmic trend [8].

The major effect in limiting thermal conductivity in 1D channels, however, seems to be boundary scattering [9, 24, 62]. Two orders of magnitude reduction in thermal conductivity has been reported for several low-dimensional materials due to roughness compared to the pristine materials, which significantly improves their thermoelectric properties [14, 62, 63]. Specifically, with regard to GNRs, studies concluded that edge roughness in GNRs can indeed reduce the thermal conductivity by up to 2 orders of magnitude, depending on the assumptions made about the roughness amplitude and the autocorrelation length.

The phonon spectrum of ultranarrow GNRs and 1D channels in general, however, consists of various phonon modes and polarizations, which react differently in the presence of disorder (i.e., LER) and exhibit different MFPs and localization lengths (LLs). To understand phonon transport at low dimensions under disorder, one needs to examine how phonon modes of all different frequencies in the entire phonon spectrum behave [29]. Understanding of the changes that the phonon modes undergo in different parts of the spectrum under strong confinement, and how these changes affect thermal transport in the presence of the LER, is also imperative in designing nanomaterials with well-controlled thermal properties. Previously, a study on thermal transport in 1D Si nanowires indicated that LER scattering affects thermal conductivity by introducing band mismatch in the optical region of the spectrum [24]. Other works attribute the reduction in thermal conductance to phonon localization and the appearance of nonpropagating modes [9, 23, 64, 65], while our work shows that all of these effects are actually present [29].

In this chapter we theoretically describe in detail the effect of the LER and confinement in phonon transport in ultranarrow AGNRs for the phonon modes of the entire energy spectrum independently. The basic conclusions of this study can be applied generically to all 1D systems. We employ the NEGF method [66], which can take into account the exact geometry of the roughness without any underlying assumptions, while we describe the phonon spectrum atomistically using force constants. We show that in the presence of the LER, all behaviors, that is, band mismatch, localization, ballisiticity, and diffusion, appear and all play a role in determining the overall thermal conductivity and its reduction under disorder. However, each effect applies to different parts of the spectrum, and each has different geometric dependence on the specific channel length and width. The chapter is organized as follows: In Section 8.2 we describe the models and methods we employ to calculate the phonon spectrum and phonon transport. In Sections 8.3 and 8.4 we present the results on the influence of the LER on phonon transmission in different parts of the phonon spectrum. Section 8.5 discusses the effect of edge roughness and GNR width on thermal conductance. In

Section 8.6 we extract the MFP and LL for GNR channels and show how different parts of the spectrum become localized at different channel lengths. Section 8.7 discusses the effects of disorder and confinement on thermal conductivity, and finally Section 8.8 summarizes and concludes the work.

## 8.2 Methods

### 8.2.1 Phonon Dispersion

Among all the models used to describe the phonon bands, such as first-principle models [67, 68], the valence force field (VFF) method [69, 70], and the force constant method (FCM), the FCM has the lowest computation time requirements. In this model, the dynamics of atoms are simply described by a few force springs connecting an atom to its surroundings up to given numbers of neighbors. The FCM uses a small set of empirical fitting parameters and can be easily calibrated to experimental measurements. Despite its simplicity, it can usually provide accurate and transferable results [71, 72].

Under the harmonic approximation, the motion of atoms can be described by a dynamical matrix as

$$D=[D_{3\times3}^{(ij)}]=\left[\frac{1}{\sqrt{M_i M_j}}\begin{cases} D_{ij} & i\neq j \\ -\sum_{l\neq i} D_{il} & i=j \end{cases}\right], \tag{8.1}$$

Where $M_{i,j}$ is the atomic mass of the *i*th and *j*th carbon atoms (in this case all atoms have the same mass) and the dynamical matrix component between atoms *i* and *j* is given by

$$D_{ij}=\begin{bmatrix} D_{xx}^{ij} & D_{xy}^{ij} & D_{xz}^{ij} \\ D_{yx}^{ij} & D_{yy}^{ij} & D_{yz}^{ij} \\ D_{zx}^{ij} & D_{zy}^{ij} & D_{zz}^{ij} \end{bmatrix}, \tag{8.2}$$

where

$$D_{mn}^{ij}=\frac{\partial^2 U}{\partial r_m^i \partial r_n^j}, \quad i,j\in N_A, \text{and} \quad m,n\in[x,y,z] \tag{8.3}$$

is the second derivative of the potential energy ($U$) after atoms $i$ and $j$ are slightly displaced along the $m$ axis and the $n$ axis ($\partial r_m^i$ and $\partial r_n^j$), respectively.

For setting up the dynamical matrix component between the $i$th and the $j$th carbon atoms, which are the $N$th nearest neighbors of each other, we use the FCM, involving interactions up to the fourth nearest neighbor [73], as shown in Fig. 8.1a. The force constant tensor is given by

$$K_0^{(ij)} = \begin{bmatrix} \phi_{\mathrm{r}}^{(N)} & 0 & 0 \\ 0 & \phi_{\mathrm{ti}}^{(N)} & 0 \\ 0 & 0 & \phi_{\mathrm{to}}^{(N)} \end{bmatrix}, \tag{8.4}$$

where $\phi_{\mathrm{r}}^{(N)}$, $\phi_{\mathrm{ti}}^{(N)}$, and $\phi_{\mathrm{to}}^{(N)}$ are the radial, the in-plane transverse, and the out-of-plane transverse components, respectively. The force constant fitting parameters are taken from Ref. [74] and are presented in Table 8.1. There are 12 fitting parameters that determine the force constants, which are extracted from fitting to experiments.

**Table 8.1** The fitting parameters of the force constant tensor in N/m for the four nearest neighbors (NN)

| | | Fitting parameters | | |
|---|---|---|---|---|
| | | $\phi_{\mathrm{r}}$ | $\phi_{\mathrm{ti}}$ | $\phi_{\mathrm{to}}$ |
| | 1 | 365 | 245 | 98.2 |
| NN | 2 | 88 | −32.3 | −4 |
| | 3 | 30 | −52.5 | 1.5 |
| | 4 | −19.2 | 22.9 | −5.8 |

*Source*: Ref. [74]

The 3 × 3 components of the dynamical matrix are then computed as

$$D_{ij} = U_m^{-1} K_0^{(ij)} U_m, \tag{8.5}$$

where $U_m$ is a unitary rotation matrix defined as:

$$U_m = \begin{bmatrix} \cos\theta_{ij} & \sin\theta_{ij} & 0 \\ -\sin\theta_{ij} & \cos\theta_{ij} & 0 \\ 0 & 0 & 1 \end{bmatrix} \tag{8.6}$$

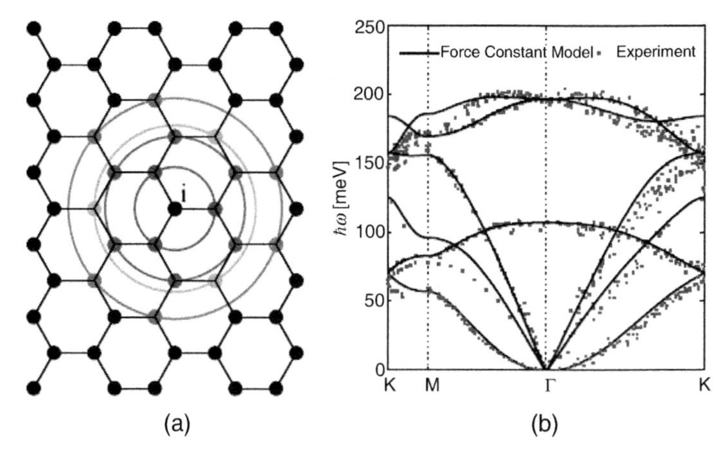


**Figure 8.1** (a) Schematic representation of the nearest neighbors of the *i*th carbon atom. Up to four nearest neighbors are included. (b) Phononic band structure of graphene (solid) evaluated using the fourth nearest-neighbor FCM. Experimental results (dots) are taken from Refs. [71, 75]. Adapted from Ref. [73]. With permission of Springer.

Assuming the graphene sheet is located in the *x*-*y* plane, $\theta_{ij}$ represents the angle between the *x* axis and the bond between the *i*th and *j*th carbon atoms.

The phonon dispersion can be computed by solving the following eigenvalue problem:

$$[D+\sum_{l} D_l \exp(i\vec{q}\cdot\Delta\vec{R})]\psi(\vec{q})=\omega^2(\vec{q})\psi(\vec{q}), \tag{8.7}$$

where $D_l$ is the dynamical matrix representing the interaction between the unit cell and its neighboring unit cells separated by $\Delta\vec{R}$ and $\psi(\vec{q})$ is the phonon mode eigenfunction at wave vector $\vec{q}$.

The evaluated phononic band structure of graphene is shown in Fig. 8.1b. It accurately reproduces the phonon dispersion of graphene. To confirm this, we present the experimental phonon band structure results from Refs. [71, 75]. As expected, our calculations are in good agreement with the experimental data, especially for the low phonon frequencies, which are the most important ones in determining thermal conductivity.

### 8.2.2 Phonon Dispersion Features

Figures 8.2a and 8.2b show typical dispersion relations for GNR channels of widths $W$ = 5 nm and $W$ = 1 nm, respectively. The $W$ = 1 nm case, as we show below, resembles purely 1D features, whereas at a width of $W$ = 5 nm the dispersion diverts towards 2D (although the dispersions in both cases are 1D). These two sizes are computationally manageable, and a comparison between their transport properties allows a comparison between 1D and less confined, "toward 2D," phonon transport. The colormap in Fig. 8.2 shows the contribution of each phonon state to the ballistic thermal conductance at room temperature.

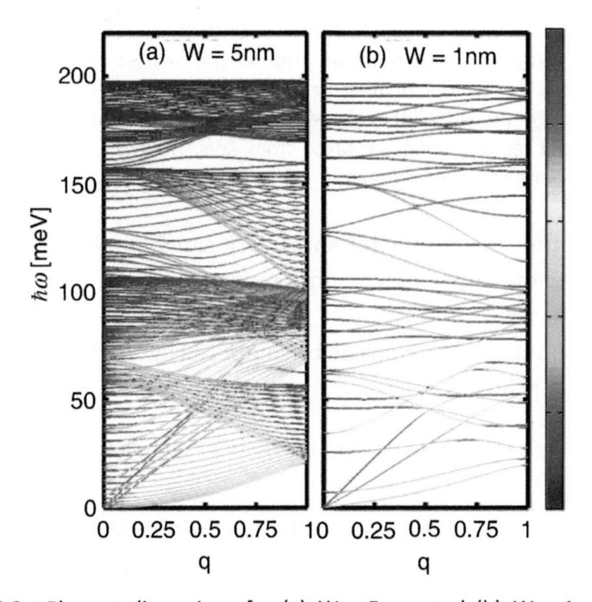


**Figure 8.2** Phonon dispersions for (a) $W$ = 5 nm and (b) $W$ = 1 nm wide armchair nanoribbons. As the width is decreased, the number of phonon modes is also reduced. The colormap shows the contribution of each phonon state to the total ballistic thermal conductance (red: largest contribution; blue: smallest contribution). Adapted figure with permission from Ref. [29]. Copyright (2015) by the American Physical Society.

To analyze the observed features of the GNR phonon dispersions, let us first consider graphene phonon dispersion. In graphene, there are six phonon modes, three acoustic and three optical modes [74]. The highest-frequency acoustic mode is the longitudinal acoustic

(LA) mode, the next one is the in-plane transverse acoustic (TA) mode, and lowest-frequency mode is the out-of-plane acoustic (ZA) mode. The latter is recently shown to make the largest contribution to the thermal conductivity of graphene [4, 5, 76–78]. The highest-frequency optical mode is the longitudinal optical (LO), followed by the in-plane transverse optical (TO), and the lowest is the out-of-plane optical (ZO) [43, 78]. The LA mode of the GNRs shown in Fig. 8.2 is the corresponding LA mode of graphene with a group velocity $v_s$ = 19.8 km/s. The LA and TA modes are linear at low frequencies and extend up to $E \approx 0.16$ eV and $E \approx 0.14$ eV, respectively. The ZA mode is quadratic for low frequencies and extends up to $E \approx 0.07$ eV. At the higher part of their energy region, the acoustic modes become relatively "flat." The ZO modes extend from $E \approx 0.7$–0.11 eV, whereas the LO and TO modes are located at higher energies, from $E \approx 0.16$–0.2 eV. The relatively flat mode regions at energies $E \approx 0.07$–0.11 eV consist of ZO modes, in addition to the dispersive LA and TA modes [43]. The less dispersive modes located from $E \approx 0.11$–0.16 eV are the flat parts of the LA and TA modes.

Three main observations on the phonon band structure can be made as the width is reduced, that is, between Fig. 8.2a and Fig. 8.2b:

- The optical and quasi-acoustic modes (which are nothing else but folded acoustic branches of the host material [79]) show strong confinement dependence [80]. The number of modes depends on the number of atoms within the unit cell. As the width is reduced from $W$ = 5 nm (Fig. 8.2a) to $W$ = 1 nm (Fig. 8.2b), the number of modes in these regions is also reduced.
- The number of acoustic modes remains intact, and they carry a much larger portion of the heat (as indicated by their red coloring in Figs. 8.2a and 8.2b).
- Small bandgaps appear in some regions in the band structure, especially in regions around the interface between the flat optical modes and the more dispersive quasi-acoustic modes (primarily around $\hbar\omega \approx 0.16$ eV and secondly around $\hbar\omega \approx$ 0.11 eV and $\hbar\omega \approx 0.07$ eV). In addition, large regions in the phononic $(\hbar\omega, q)$ space, especially in the quasi-acoustic band regions, become 'empty' of modes (sparse), where for rather extensive energy and momentum intervals no phonon states exist.

### 8.2.3 Phonon Transport within NEGF

The FCM can be coupled to NEGF for the calculation of the coherent phonon transmission function of the GNR. The NEGF method is appropriate for studies of phonon transport in geometries with disorder because the exact geometry is included in the construction of the dynamical matrix. Employing an atomistic approach that considers the discrete nature of the LER and accurately models its impact on phonon modes is essential for the analysis of thermal properties of narrow GNRs (with $W < 20$ nm). The method considers the wave nature of phonons, rather than their particle description, and all interference and localization effects, which could be important in low-dimensional channels, are captured. In addition, it is most appropriate for the purposes of this study, which investigates the influence of the LER for phonons of different frequencies of the spectrum, as NEGF computes the energy resolved phonon transmission function. The system geometry consists of two semi-infinite contacts made of pristine GNRs, surrounding the channel in which we introduce the LER. Green's function is given by

$$G(E) = [E^2 I - D - \Sigma_1 - \Sigma_2]^{-1}, \tag{8.8}$$

where $D$ is device dynamical matrix and $E = \hbar\omega$ is the phonon energy. The contact self-energy matrices $\Sigma_{1,2}$ are calculated using the Sancho–Rubio iterative scheme. The transmission probability through the channel can be obtained using the relation

$$T_{\text{ph}}(\omega) = \text{Trace}[\Gamma_1 G \Gamma_2 G^+], \tag{8.9}$$

where $\Gamma_1$ and $\Gamma_2$ are the broadening functions of the two contacts defined as $\Gamma_{1,2} = i\left[\Sigma_{1,2} - \Sigma_{1,2}^+\right]$. The thermal conductance can then be calculated in the framework of the Landauer formalism as

$$K_l = \frac{1}{2\pi\hbar}\int_0^\infty T_{\text{ph}}(\omega)\hbar\omega\left(\frac{\partial n(\omega)}{\partial T}\right)d(\hbar\omega), \tag{8.10}$$

where $n(\omega)$ is the Bose–Einstein distribution and $T$ is the temperature. In this work we consider room temperature $T = 300$ K. At room temperature and under ballistic conditions the function inside the integral spans the entire energy spectrum [58, 81], which allows phonons of all energies to contribute to the thermal conductance.

# 8.3 Influence of Roughness on Phonon Transport

## 8.3.1 Influence of Roughness on Phonon Transmission

We then investigate phonon transport in these low-dimensional GNRs in the presence of disorder. At such small ribbon widths with rough edges, the edge phonon scattering is the dominant scattering mechanism [25]. For this, we simulate rough GNR channels of width $W = 5$ nm (relatively wide) down to $W = 1$ nm (purely 1D) and examine the phonon transmission across the phonon energy spectrum as the length of the GNR increases (i.e., as the effective disorder increases). We construct the LER geometry by adding carbon atoms to or subtracting carbon atoms from the edges of the pristine GNR according to the exponential autocorrelation function

$$R(x) = \Delta W^2 \exp\left(-\frac{|x|}{\Delta L}\right), \tag{8.11}$$

where $\Delta W$ is the root mean square of the roughness amplitude and $\Delta L$ is the roughness correlation length [26]. The Fourier transform of the autocorrelation is the power spectrum of the roughness. The real space representation of the LER is achieved by adding a random phase to the power spectrum followed by an inverse Fourier transform [26, 82]. We use $\Delta W = 0.1$ nm and $\Delta L = 2$ nm. Numerically, we construct a roughness line at the top and bottom edges of the GNR and atoms located in the outer direction of the lines are removed, whereas all regions in the inner part of the lines are filled with atoms if needed. We keep this roughness description constant in all cases. Here we only consider GNRs with armchair edges. However, we have also performed simulations for GNRs with zigzag edges and we verify that the main conclusions of this work still apply to the ZGNR case as well [29]. All the GNRs are assumed to be suspended, and the edge atoms are allowed to move freely, vibrating in harmony with the other atoms according to the lattice wave/dynamics. Therefore, the effective disorder in the channels we simulate increases as (i) the channel length is increased or (ii) the channel width is reduced.

To provide an indication of the computational cost, we note that the largest nanoribbon we consider, of width $W = 5$ nm and channel length $L$ = 500 nm, consists of nearly 100,000 atoms. To describe the motion of each atoms a 3 × 3 matrix is needed. The resulting dynamic matrix and Green's functions at each energy point are matrices with a size of 300,000 × 300,000. To simulate the phonon transmission of this structure, it takes several hours on a 16-core machine with 128 GB of memory. Since the roughness produces a random edge topology, all results we present are extracted by averaging 50 different realizations of roughened channels. Note that what we present here is an investigation of the effects of disorder on the phonon transport of 1D channels. Using the same computational resources we could scale the GNR width even up to $W$ = 50 nm, but in that case we would have to reduce the length down to $L$ = 50–100 nm. At such wide and short channels, though, the effects of edge roughness would have been minimized.

Figure 8.3 shows the transmission function of the phonon spectrum as a function of energy for the GNR with width $W$ = 5 nm (Fig. 8.3a) and for the ultranarrow GNR of width $W$ = 1 nm (Fig. 8.3b). The figure shows transmissions of channels with rough edges and various lengths. The dashed black lines indicate the ballistic transmission of the GNRs with perfect edges. The transmissions of GNRs with lengths $L$ = 5 nm (blue line), $L$ = 40 nm (red line), $L$ = 100 nm (green line), and $L$ = 500 nm (solid black line) are plotted.

In the case of ballistic transport, the transmission is significant in the entire energy spectrum, and thus the whole spectrum contributes to thermal conductance for both wide and narrow GNRs [26]. Of particular note is the sharp transmission peak in the high-energy optical modes in the case of the wide GNR in Fig. 8.3a, which originates from their large number rather than their group velocity, which is low. The LER reduces the transmission function significantly and, in particular, around energies $E$ = 0.06–0.07 eV, $E$ = 0.11–0.14 eV, and $E$ = 0.16–0.17 eV. This group of energy regions, for which the transmission is strongly reduced, is of low-density (but also dispersive) modes. In particular, the latter energy region is the one around the boundary between flat and dispersive modes, exactly above the energy at which the LA mode ends, and is a region with particularly low mode density. A surviving contribution to the transmission is evident around energies $E$ = 0–0.05 eV (acoustic phonons), $E$ = 0.08–0.11 eV (a mixture of LA, TA, and ZO modes), and $E$ = 0.17–0.2 eV (optical phonons), even for the longer-length GNRs.

It is evident from this that the low-group-velocity optical modes contribute significantly to transmission due to their large density, even in the presence of roughness.

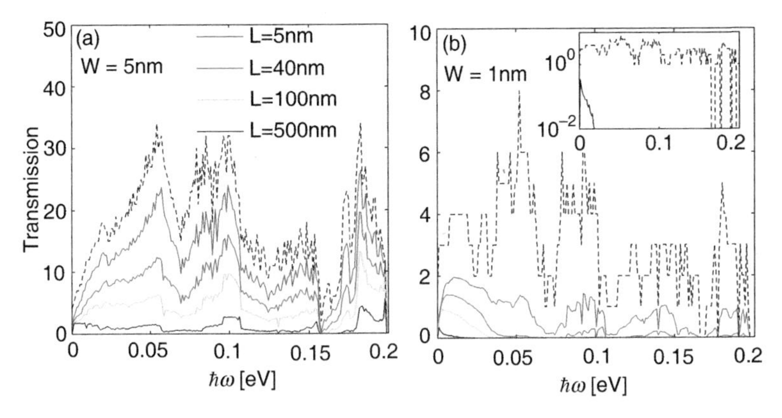


**Figure 8.3** Transmission function versus energy for rough-edge GNRs of width (a) $W$ = 5 nm and (b) $W$ = 1 nm. Nanoribbon lengths of $L$ = 5 nm (blue lines), $L$ = 40 nm (red lines), $L$ = 100 nm (green lines), and $L$ = 500 nm (black lines) are considered. The ballistic transmissions (pristine, nonroughened ribbons) are depicted in dashed black lines. Adapted figure with permission from Ref. [29]. Copyright (2015) by the American Physical Society.

The corresponding transmission functions for the narrower GNR, with the width $W$ = 1 nm, shown in Fig. 8.3b, undergo much stronger reductions with the LER compared to the wider GNRs of the same length. Since we keep the roughness amplitude the same in all cases, reducing the width essentially increases the effective disorder. The reduction is much stronger in the entire energy spectrum, in particular around the low-density-mode energy regions ($E$ = 0.06–0.07 eV, $E$ = 0.11–0.14 eV, and $E$ = 0.16–0.17 eV, as mentioned before), where the transmission is diminished. Dominating thermal conductance in the ultranarrow GNR case, especially when the length of the channel is increased above $L$ > 40 nm, are the low-energy, low-wave-vector acoustic modes (solid black line in Fig. 8.3b). This is clearly indicated in the inset of Fig. 8.3b, which shows on a logarithmic scale the transmission of the ballistic GNR channel and the transmission of the rough-edge GNR channel with $L$ = 1 μm and $W$ = 1 nm. Clearly, only the transmission in the low-energy region survives.

### 8.3.2 Influence of Roughness on Different Phonon Modes

To illustrate the distinctly different behavior of the various phonon modes in the presence of the LER, Fig. 8.4 shows the transmission at certain phonon frequencies as a function of the channel length $L$. Figures 8.4a, 8.4b, and 8.4c show results for the $W = 5$ nm, $W = 3$ nm, and $W = 1$ nm GNRs, respectively.

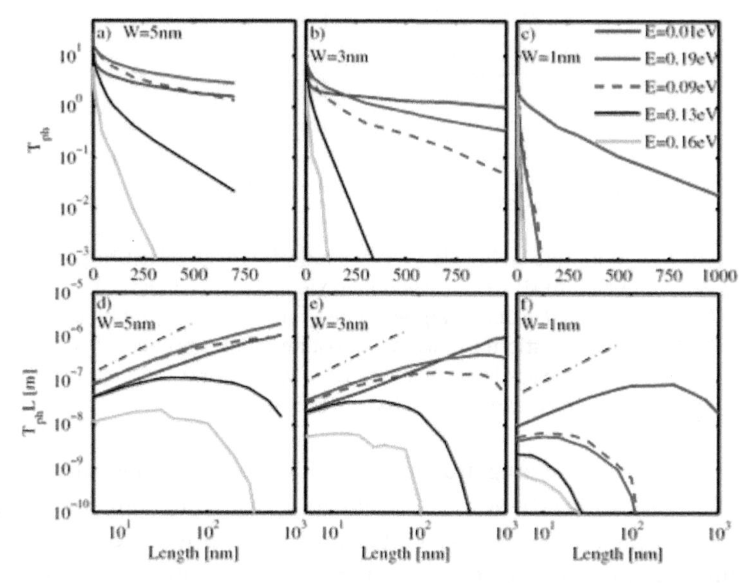


**Figure 8.4** (a–c) Phonon transmission of rough nanoribbons of widths $W$ = 5 nm (a), $W$ = 3 nm (b), and $W$ = 1 nm (c) for specific energies versus channel length. Energies $E$ = 0.01 eV (blue lines) correspond to the acoustic branches. $E$ = 0.19 eV and $E$ = 0.09 eV (solid red and dashed red lines, respectively) correspond to regions of the spectrum where the bands are numerous, but mostly flat. $E$ = 0.16 eV (green line) corresponds to a region of the spectrum at the interface between dispersive and flat bands, in which narrow bandgaps are formed as the width is reduced. $E$ = 0.13 eV (black line) corresponds to a spectrum region where dispersive bands exist, but as the width is reduced they are reduced in number and in addition narrow bandgaps form. (d, e) The phonon transmission times the channel length $T \times L$ for the same situations as in (a–c). Adapted figure with permission from Ref. [29]. Copyright (2015) by the American Physical Society.

We concentrate on four different phonon categories and pick a specific phonon energy within the energy region of these categories.

These are (i) acoustic phonons ($E$ = 0.01 eV, blue lines), (ii) optical, flat dispersion phonons ($E$ = 0.19 eV, solid red lines; and $E$ = 0.09 eV, dashed red lines), (iii) quasi-acoustic, dispersive phonon modes ($E$ = 0.13 eV, black lines), and (iv) regions of very low mode densities, in which confinement can even result in narrow bandgaps ($E$ = 0.16 eV, green lines). For all energy cases and for all GNR widths, the transmission drops with increasing channel length and reducing width. The drop, however, differs significantly for each different phonon energy case. The drop in the transmission of the acoustic modes (blue lines) is relatively weak and can be understood from the fact that they are composed of LA modes with long wave vectors [10, 11]. These modes are very weakly affected by defects, and this is the case for both wider and ultranarrow GNRs. For example, Scuracchio et al. [83] have also indicated that these modes are only weakly affected by atomic vacancies, and Huang et al. [84] reached very similar conclusions in the presence of dislocation defects in GNRs. The optical modes (solid red and dashed red lines) have a much stronger dependence on the GNR width. For the wider channel (Fig. 8.4a), their transmission is even larger compared to the acoustic modes independent of channel length. As the width is reduced, their transmission drops with increasing length, especially in the case of the ultranarrow $W$ = 1 nm, channel. In the case of the quasi-acoustic modes (black lines), a large drop in the transmission is observed as the channel length increases. Even stronger is the drop in the transmission of the very low-density-mode regions (green lines). In the following sections, we provide explanations regarding this behavior.

### 8.3.3 Ballistic, Diffusive, and Localized Phonon Modes

In recent experiments in graphene and carbon nanotubes it was shown that thermal transport could deviate from Fourier's law and exhibit semiballistic behavior [6, 8]. Since each phonon mode responds differently to disorder, it is essential to investigate the regions of operation of the different modes and identify the ones that contribute to the semiballistic behavior. Figures 8.4d, 8.4e, and 8.4f, show the product of the transmission times the length of the channel ($T \times L$) versus channel length $L$ for the same channels and phonon modes as in Figs. 8.4a, 8.4b, and 8.4c, respectively. In the case of ballistic transport, the $T \times L$ product increases linearly. In the case of diffusive transport it remains constant. In the case of subdiffusive transport the product reduces with length [85–87], and for localized transport, the product drops exponentially. From Figs. 8.4d and 8.4e, it can be observed that for the wider GNR channels, the acoustic modes (blue lines) are semiballistic, even for channel widths $W$ = 3 nm and lengths up to $L$ = 1 μm. For the ultranarrow $W$ = 1 nm, GNRs (Fig. 8.4f), the acoustic modes reach the diffusive regime at around lengths of $L \approx 200$ nm and get into the localized regime for lengths larger than $L \approx 700$ nm. Interestingly, a similar trend is observed for the optical modes (red lines) as well. For GNR widths $W$ = 5 nm (Fig. 8.4d) and $W$ = 3 nm (Fig. 8.4e), they indicate a semiballistic behavior even up to channel lengths of hundreds of nanometers. In the $W$ = 1 nm case, though, the optical modes reach the localization regime at lengths well below $L \approx 100$ nm. The behavior of the quasi-acoustic modes (black lines), on the other hand, is very different. These modes enter the diffusive regime at much shorter channel lengths compared to the acoustic and optical modes. They even enter the localization regime after $L \approx 300$ nm for the $W$ = 5 nm GNRs, after $L \approx$ 100 nm for the $W$ = 3 nm GNRs, and just after $L \approx 10$ nm for the $W$ = 1 nm GNRs. This is quite intriguing since these are dispersive modes with much higher group velocities than the optical modes. The strongest reduction in transmission, however, is observed for the energy regions of low mode density (green lines). For these modes, the transmission is completely diminished after channel lengths of $L \approx 100$ nm in the case of the wider channels and after $L \approx 10$ nm in the case of the ultranarrow channel.

An important point to clarify here is that the diffusive regime in our calculations is caused by edge roughness and not phonon-phonon interactions, which are not included in the simulations. Including the full unharmonicity of the material within the NEGF would have been a much more computationally intensive effort. However, edge roughness under ultranarrow channels could have a much stronger effect compared to phonon-phonon interactions and the thermal conductance could reach diffusive, even localization behavior even before phonon-phonon interaction becomes important. This depends on how much smaller the MFP of edge disorder is compared to the MFP of phonon-phonon interaction (which is in the hundreds of nanometers in graphene). A discussion on the characteristic length scales is presented further in Section 8.6.

## 8.4 Transmission Effects in Width-Modulated GNRs

In Figs. 8.5 and 8.6 we provide explanations for the behavior of the transmission in the different phonon energy regions with channel length and width. We base our analysis on two effects that explain the behavior of the modes: (i) the change in the phonon band structure at specific energies under the influence of roughness and (ii) the corresponding change under the influence of geometrical confinement. We demonstrate that increasing effective roughness has a similar effect as increasing confinement. For example, regions in the phonon spectrum that become sparse of modes due to confinement tend to more easily form effective bandgaps in the presence of roughness as well, driving the transmission into localization. Figure 8.5 discusses the effect of confinement on specific energy regions of the band structure, whereas Fig. 8.6 the effect of roughness, specifically on the sparse mode regions.

In Fig. 8.5 we consider the $W$ = 1 nm GNR and the following situation: We simulate the phonon modes and transmission for the ultranarrow GNR of width $W$ = 1.1 nm, a GNR of width $W$ = 0.74 nm, and a GNR whose width is periodically modulated along its length (rather than randomly as in the case of rough channels), as shown in Fig. 8.5f (lower part in blue). In this case, we can isolate the influence of roughness on the band structure. The left parts of the panels of Fig. 8.5 show the phonon band structure of the three channels in the vicinity of the energies of interest. The band structure for the wide channel is shown in red, for the narrow channel in green, and for the width modulated channel in blue. The corresponding parts on the right show the transmission of the three channels. Figures 8.5a to 8.5e show, respectively, results for energies around $E$ = 0.001 eV (low-frequency acoustic modes), $E$ = 0.09 eV and $E$ = 0.19 eV (optical modes), $E$ = 0.16 eV (low-density-region modes), and $E$ = 0.13 eV (quasi-acoustic modes).

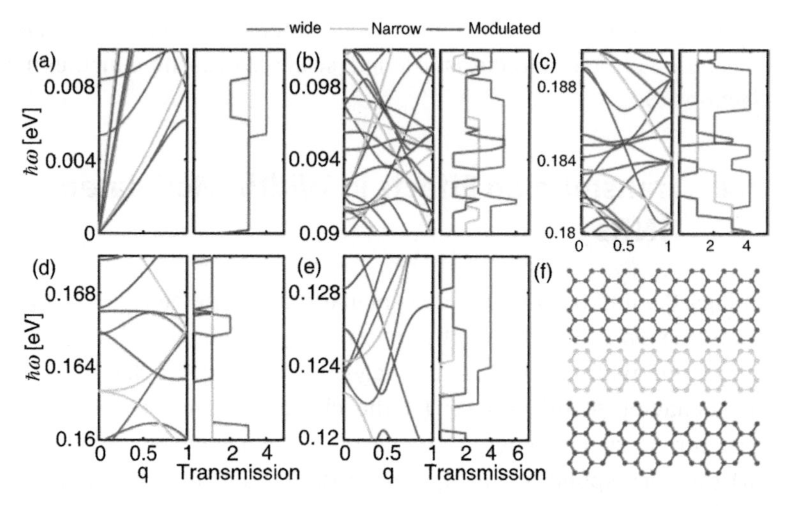


**Figure 8.5** Phonon dispersion and transmission function of the $W = 1$ nm GNR under three different situations as shown in panel (f): A slightly wider channel of $W = 1.11$ nm (red), a slightly narrower channel of $W = 0.74$ nm (green), and a GNR whose width is periodically modulated (blue) are considered. The latter mimics a rough ribbon. Different sets of energies are shown: (a) $E = 0$ eV to $E = 0.01$ eV (acoustic modes), (b) $E = 0.09$ eV to $E = 0.1$ eV (optical modes), (c) $E = 0.18$ eV to $E = 0.19$ eV (optical modes), (d) $E = 0.16$ eV to $E = 0.17$ eV (regions between quasi-acoustic and optical modes), (e) $E = 0.12$ eV to $E = 0.13$ eV (quasi-acoustic modes), and (f) schematic of the atomistic geometries of the three nanoribbon cases. Adapted figure with permission from Ref. [29]. Copyright (2015) by the American Physical Society.

## 8.4.1 Influence of Width Modulation on Acoustic Modes

In the case of the low-frequency acoustic modes in Fig. 8.5a, the transmission of the modulated channel is dominated by the transmission of the narrow region. In a small energy range a band mismatch is observed around the edge of the Brillouin zone, and the transmission is further reduced. In general, however, the reduction in transmission is relatively weak, which explains the fact that these modes behave semiballistically, especially as the energy and wave vector approach zero.

## 8.4.2 Influence of Width Modulation on Optical Modes

In the case of optical, flat dispersion modes at energies $E \approx 0.09$ eV and $E \approx 0.19$ eV, it is evident from Fig. 8.5b and Fig. 8.5c that the reduction in the transmission due to width modulation (or roughness) originates from a band mismatch between the narrow and wider GNRs. The transmission of the width-modulated GNR is actually lower compared to the transmissions of both the wide and narrow GNRs. For this $W = 1$ nm GNR, the density of optical modes is rather low and the mismatch that is created under width modulation along the length of the channel can be significant, which degrades the transmission.

### 8.4.3 Influence of Width Modulation on Low-Density-Mode Regions

Figure 8.5d shows the width-modulated results for the low-density-mode energy regions at energies $E \approx 0.16$ eV. As in the case of optical modes, a strong mismatch can be observed between the bands of the width-modulated GNR and the bands of both the wide and narrow GNRs. The mismatch, however, is much larger, at a degree where energy bandgaps are formed in the transmission function (Fig. 8.5d, right panel). Note that small bandgaps are also formed even in the uniform channels under strong confinement around this energy, which further increases the band mismatch in the presence of the LER. The combination of bandgap formation and band mismatch justifies the drastic transmission drop for this particular energy region as the channel length increases (see, for example, Fig. 8.3, green lines).

### 8.4.4 Influence of Width Modulation on Quasi-Acoustic Modes

Moving along to the case of the quasi-acoustic modes of energy $E \approx 0.13$ eV, shown in Fig. 8.5e, it is evident that the bands of the width-modulated GNR can look quite different compared to the bands of the wide or narrow GNRs. Some mode mismatch can be observed, which reduces the transmission even down to zero in certain parts of the spectrum. This, however, only partially explains why the drop with channel length, shown in Fig. 8.4 (black lines), is so strong, that is, it is much stronger compared to the drop in the optical modes at energy $E \approx 0.09$ eV or $E \approx 0.19$ eV.

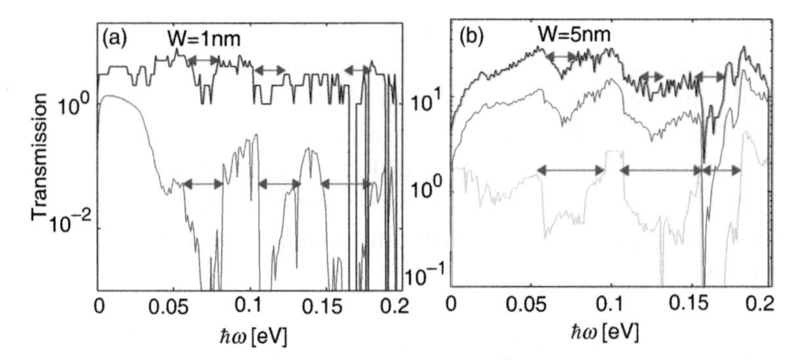


**Figure 8.6** Transmission function versus energy on a logarithmic scale for rough-edge GNRs of width (a) $W$ = 1 nm and (b) $W$ = 5 nm. The ballistic transmission (pristine GNRs, nonroughened ribbons) is depicted by the black lines. The transmission of nanoribbons with length $L$ = 40 nm is shown by the red lines. In (b) the transmission of the GNR with length $L$ = 500 nm is also shown in green. Adapted figure with permission from Ref. [29]. Copyright (2015) by the American Physical Society.

The reason why the quasi-acoustic modes behave so drastically different compared to the optical modes can be explained by looking at their behavior under confinement. Figure 8.2 shows that under confinement, the number of modes in these energy regions ($E \approx 0.07$ and $E \approx 0.13$) is reduced significantly, making these regions look almost empty of modes. In the presence of the LER in a real geometry, the sparsity of the modes makes these particular energy regions more susceptible to the formation of effective bandgaps by increasing the band mismatch. Such an event is not the case for the optical modes for the geometries we examine. The effective transmission bandgap formation is demonstrated in the transmission functions shown on a logarithmic scale in Fig. 8.6. Figure 8.6a shows the logarithmic transmission of the $W$ = 1 nm GNR under ballistic (pristine channel) conditions (black line) and under the LER when the channel length is $L$ = 40 nm (red line). It is evident that for energies $E \approx 0.07$ eV and $E \approx 0.13$ eV large effective bandgaps form as indicated by the arrows, which become wider as the channel length increases even further (not shown). Figure 8.6b shows the same transmissions for the $W$ = 5 nm GNR, but in this case we plot the transmission for the GNR with $L$ = 500 nm as well (green line). For short channels, the transmission is not significantly disturbed, but for the longer channels, bandgaps similar to the ones of the $W$ = 1 nm GNR of Fig. 8.6a form at $E \approx$ 0.07 eV and $E \approx 0.13$ eV, as also indicated by the arrows. Notice the even larger bandgap formation at energies $E \approx 0.16$ eV. This clearly indicates that the energy regions that become sparse of modes under confinement are very susceptible to roughness in less confined geometries as well, which suggests that the influence of confinement has similar features in the transmission as the effect of roughness.

### 8.4.5 General Discussion of Width-Modulated Features

The behavior described before should hold for any sparse mode energy regions. Note, for example, that gaps do not form in the regions of the flat optical modes and the transmission does not degrade as much. Under strong confinement, however, the flat optical mode regions become sparser and in extreme cases begin to look like the low-density regions as well. Under these conditions, they could also be subject to the effect we describe above. In this context, the thermal conductivity is a function of the width-dependent phonon spectra [25], for which the LER could either further increase the band mismatch or form effective transport bandgaps.

We mention here that as in the case of electronic transport, the chirality of GNRs, that is, AGNRs or ZGNRs, can provide anisotropy in phonon transport behavior (although smaller compared to electronic transport anisotropy) [88]. In Ref. [89], for example, using the phonon Boltzmann transport equation (BTE), it was shown that the amount of anisotropy between AGNR and ZGNR ribbons can be significant and increases as the ribbon width decreases and as the roughness amplitude increases. However, the bands and the transmission of the ZGNR change under confinement and roughness, indicating very similar qualitative behavior as for the AGNRs [29]. Thus, an important message we convey here is the fact that just by looking at how the phonon band structure behaves under confinement, and at its low-dimensional dispersion features, one can provide an indication of how the modes will behave under edge roughness. We do not focus specifically on the details of the GNR dispersion itself, but we rather provide general low-dimensional phonon transport features. Qualitatively, the behavior we describe should hold for other low-dimensional materials but could also be relevant to graphene ribbon phonon dispersions extracted through density functional theory (DFT) calculations (using local density approximation [LDA], generalized gradient approximation [GGA], or

GW, which can produce slightly different dispersions with respect to each other) and might also produce slightly different dispersions compared to the ones obtained using the FCM we employ here. Indeed, several works have investigated the phonon dispersions and phonon localization in GNRs using DFT calculations [90–93], with mainly similar observations.

Of course, the benefit of the FCM is that it can be combined with NEGF and describe long rough channels with hundreds of thousands of atoms, something computationally impossible with ab initio methods. For example, previously we have shown that the FCM (as a semiempirical method with fitting parameters) can correctly regenerate the band structure of graphene, obtained from first-principle calculations [73]. Furthermore, we have shown that by employing this approach for a relative roughness between ~0.5% and ~5% of the ribbon's width, very good agreement with the experimental data for GNRs with widths up to ~15 nm can be achieved [26].

## 8.5 Thermal Conductance

We next consider the thermal conductance of the GNRs at $T = 300$ K in the presence of the LER. We consider channels of different widths and lengths, as shown in Fig. 8.7a. The thermal conductance drops as the channel length increases, and the reduction rate, if compared to Fig. 8.4a–c, follows the reduction in the transmission of the dominant modes. For the wider GNRs, the reduction rate is smaller, as the transmission of the dominant acoustic and optical bands is affected only slightly. As the width is reduced down to the ultranarrow $W = 1$ nm, the thermal conductance drops faster with channel length (dotted blue line).

Interestingly, the product of thermal conductance times channel length $K \times L$ in Fig. 8.7b shows, however, that only the wider channel with $W = 5$ nm operates in the quasi-ballistic regime ($K \times L$ continues to increase even up to channel lengths of $L = 750$ nm). The channels with widths $W = 4$ nm, 3 nm, and 2 nm operate in the diffusive regime (caused by roughness, not phonon-phonon scattering) for channel lengths beyond $L = 500$ nm ($K \times L$ saturates to a constant value). The ultranarrow $W = 1$ nm channel, on the other hand, for channel lengths $L > 300$ nm enters the localization regime ($K \times L$

decreases—see inset of Fig. 8.7b). In either channel case, modes exist that are ballistic, diffusive, or localized, as discussed above. The overall behavior at larger channel lengths, however, is dominated by the behavior of the acoustic modes (the wider GNRs have a strong contribution from the optical modes as well).

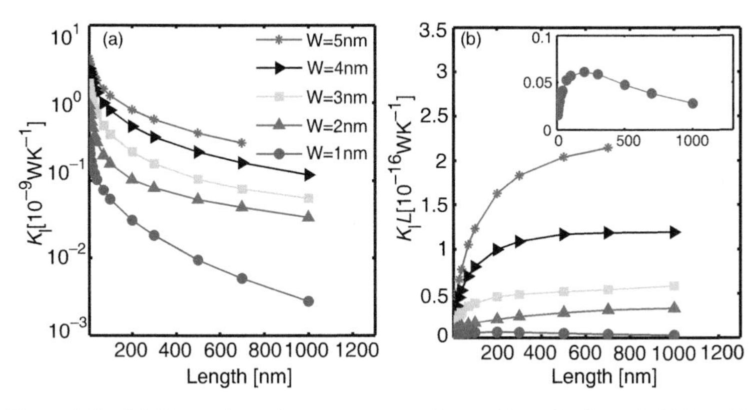


**Figure 8.7** (a) Thermal conductance versus channel length of rough GNRs with widths $W$ = 5 nm (red crosses), $W$ = 4 nm (black triangles), $W$ = 3 nm (green squares), $W$ = 2 nm (red triangles), and $W$ = 1 nm (blue circles) is shown. (b) The same channels as in (a) but the thermal conductance times the channel length $K_l \times L$ is shown. (b, inset) Zoom-in view for the $W$ = 1 nm case. Adapted figure with permission from Ref. [29]. Copyright (2015) by the American Physical Society.

The dominance of the acoustic modes is clearly illustrated in Figs. 8.8a, 8.8b, and 8.8c, which show the cumulative thermal conductance at room temperature as a function of energy for the GNRs of widths $W$ = 5 nm, $W$ = 3 nm, and $W$ = 1 nm, respectively. Results for GNRs of lengths $L$ = 5 nm (blue lines), 40 nm (red lines), 100 nm (green lines), and 500 nm (black lines) are shown. The dashed lines show the cumulative ballistic thermal conductance. In the ballistic case, independent of the GNR width, the entire spectrum contributes to thermal conductance, with the low-energy acoustic modes contributing ~50% and the high-energy optical modes ~10%, whereas the rest ~40% is contributed from phonons in the intermediate-energy region. For the roughened wider GNR with $W$ = 5 nm (Fig. 8.8a), this behavior is also independent of channel length and retained until at least $L$ = 500 nm. As the width of the GNR is reduced, as in the $W$ = 3 nm GNR case shown in Fig. 8.8b, the situation is similar, except that at larger channel lengths, the contribution of the low-energy phonons increases. In the case of longer channels, the higher-energy modes get into subdiffusion and/or localization regimes and contribute less. This results in ~80% of the heat being carried by phonons with energies below $E = 0.02$ eV. For even narrower GNRs, as the ultranarrow $W = 1$ nm GNR shown in Fig. 8.8c, the distribution shifts toward the low-energy acoustic modes at much shorter channel lengths, even as short as $L = 5$ nm (blue line). In the limit of very long and very narrow channels, that is, approaching purely 1D, all heat is carried by the very low-energy acoustic modes, whereas all higher-energy modes are driven into the localization regime [6, 58].

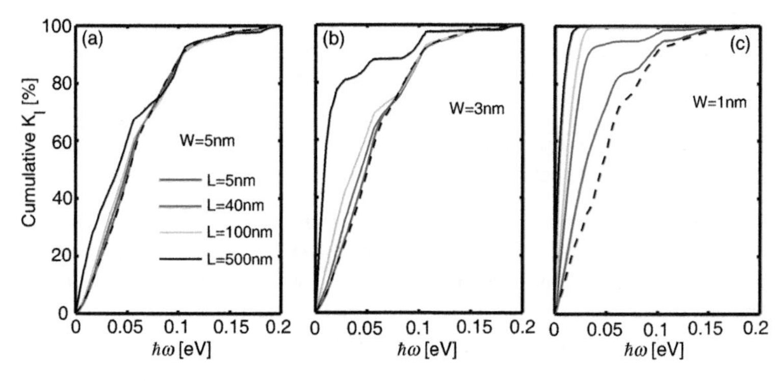


**Figure 8.8** Cumulative thermal conductance versus energy for GNR channels of different widths: (a) $W = 5$ nm, (b) $W = 3$ nm, and (c) $W = 1$ nm. For every case, the dashed line indicates the ballistic case. Channel lengths of $L = 5$ nm (blue), $L = 40$ nm (red), $L = 100$ nm (green), and $L = 500$ nm (black) are shown. Adapted figure with permission from Ref. [29]. Copyright (2015) by the American Physical Society.

## 8.6 Characteristic Scattering Length Scales

To identify the dependence of the transmission function on the channel length for the different operating regimes, we need to relate it to the MFP for scattering, $\lambda$, and the localization length (LL), $\zeta$.

### 8.6.1 Mean Free Path for Scattering

A calculation of the phonon MFP gives an estimate of the distance over which the phonons travel before they scatter (in this case off the boundary), and can provide an understanding of the thermal transport process. The LER scattering-limited transmission function $T_{\text{LRS}}(\omega)$ is related to the ballistic transmission $T_{\text{B}}(\omega)$, $\lambda(\omega)$, and the channel length $L$ by the relation [49]

$$T_{\text{ph}}(\omega)=\frac{\lambda(\omega)}{L+\lambda(\omega)}T_{\text{B}}(\omega). \tag{8.12}$$

From this, the LER-limited MFP can be extracted as

$$\lambda(\omega)=\frac{T_{\text{ph}}(\omega)L}{T_{\text{B}}(\omega)-T_{\text{ph}}(\omega)}. \tag{8.13}$$

When writing down Eq. 8.12, we assume that the system can be seen as two thermal resistances in series, the channel, and the contacts where the phonons thermalize. Thus, the MFP increases with channel length $L$ until the channel enters the diffusive regime. Strictly speaking, only then does the diffusive MFP converge and can be extracted. While this condition can be reached for short channel lengths for most phonon energies, the acoustic phonons, which carry most of the heat, have very long MFPs, beyond the channel lengths we could simulate. (To provide an indication of the computational cost, we note that a nanoribbon with a width of 5 nm and a channel length of 500 nm [the biggest we simulate] consists of nearly 100,000 atoms. To describe the motion of each atom a 3 × 3 matrix is needed. The resulting dynamic matrix and Green's functions at each energy point are matrices with a size of 300,000 × 300,000. Thus, increasing the length largely increases the computational cost). Therefore, to increase the accuracy in extracting the MFP, we use the transmission values at two different channel lengths as [24]

$$\lambda(\omega)=\frac{T_{\text{LRS},L_2}(\omega)L_2-T_{\text{LRS},L_1}(\omega)L_1}{T_{\text{LRS},L_1}(\omega)-T_{\text{LRS},L_2}(\omega)}, \tag{8.14}$$

which accounts partially for the fact that the transmission of phonons with long MFPs has not yet converged fully for the simulated channel length $L$.

Figure 8.9a shows the average diffusive phonon MFP for scattering on the rough boundaries, $\lambda(\omega)$, as a function of frequency for the channels of two different widths $W$ = 5 nm (solid red line) and $W$ = 1 nm (solid blue line). The MFP is extracted as specified by Eq. 8.14. Since each frequency region, however, enters the diffusive regime at different channel lengths, the MFP for every energy is extracted at the channel length at which the product of the transmission times length ($T \times L$ as in Fig. 8.4d–f) becomes constant, or levels out. Therefore, Fig. 8.9 considers a different channel length at all energies for both channel widths and both $L_1$ and $L_2$ taken at each instance when $T \times L$ levels out. For the wider $W = 5$ nm channel, the average (averaged over 50 geometry realizations) diffusive MFP (solid red line) varies from a few tens of nanometers up to even a few hundreds of nanometers, in agreement with other works as well [57]. It only drops to a few nanometers at energies $E \approx 0.16$ eV due to the large mismatch between the modes in this sparse mode energy region and the formation of a transport gap. For the ultranarrow $W = 1$ nm channel (solid blue line), very large MFPs on the order of several hundreds of nanometers are observed for the low-frequency phonons close to the zone center originating from the LA modes. This is similar to the phonon-phonon scattering MFP reported in other carbon nanostructures such as carbon nanotubes and graphene sheets, which is reported to be ~800 nm [30, 94–96], even in the presence of defects [34]. For slightly larger energies, that is, $E > 0.03$ eV, the MFP drops sharply to very low values, of at most a few nanometers.

An average MFP value for the entire energy range can be extracted as

$$\langle\langle\lambda\rangle\rangle = \frac{\int \lambda(\omega) T_{\text{ph}}(\omega) W_{\text{ph}}(\omega) d\omega}{\int T_{\text{ph}}(\omega) W_{\text{ph}}(\omega) d\omega}, \tag{8.15}$$

where the phonon window function $W_{\text{ph}}(\omega)$ is given by

$$W_{\text{ph}}(\omega) = \frac{3}{\pi^2}\left(\frac{\omega}{k_B T}\right)^2\left(-\frac{\partial n}{\partial \omega}\right). \tag{8.16}$$

Our calculations show that the average LER-limited diffusive MFP in the case of the narrow GNRs is $\langle\lambda\rangle \approx 30$ nm, whereas for the wider GNR of $W = 5$ nm, it largely increases to $\langle\lambda\rangle \approx 600$ nm. It should be noted that the inclusion of phonon-phonon interaction, which is neglected in this work, can result in smaller MFPs, especially for the high-energy optical modes. An accurate modeling of phonon-phonon interaction due to anharmonicities is beyond the scope of this work.

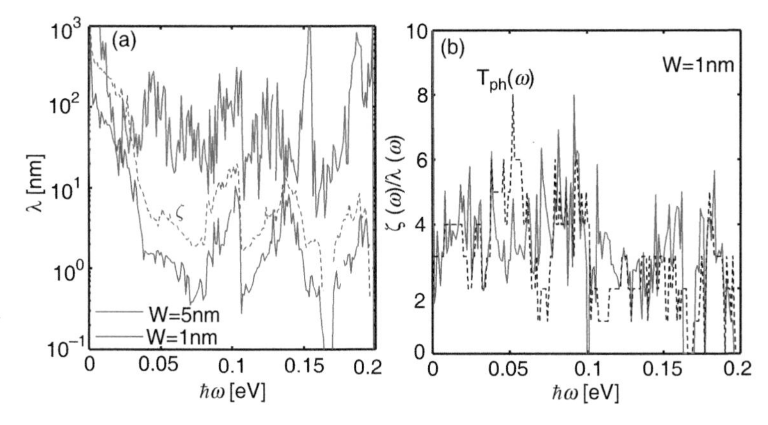


**Figure 8.9** (a) Average diffusive transport mean free path (solid) versus energy of rough GNRs of widths $W$ = 5 nm (red) and $W$ = 1 nm (blue). The MFP as a function of energy is extracted at the channel length at which the $T \times L$ product is constant; therefore, the channel length differs for each energy value. The dashed blue line shows the localization length for the $W$ = 1 nm. (b) Ratio of the localization length $\zeta(\omega)$ over the MFP $\lambda(\omega)$ for the $W$ = 1 nm rough ribbon of length $L$ = 1000 nm (blue line) and the transmission probability of the pristine $W$ = 1 nm GNR (dashed black line). Adapted figure with permission from Ref. [29]. Copyright (2015) by the American Physical Society.

## 8.6.2 Localization Length

Localization is a transport regime that dominates the flow of phonons in low-dimensional channels, and several authors have studied it. Loh et al. [97] studied the influence of vacancies in phonon transport in graphene and showed that acoustic modes tend to get localized around the vacancies, which results in a large reduction in the thermal conductivity. Isotope scattering could also cause localization of high-energy optical phonon modes in graphene [98]. Kim et al. [99] have recently also experimentally showed that phonon localization appears in graphene in the presence of graphene nanobubbles formed by noble gas atom implantation. Wang et al. [10] showed that the anisotropic behavior of thermal conductivity between zigzag- and armchair-edge GNRs originates from the stronger localization around the armchair edges. In a different work, Jiang et al. [100] studied phonon transport in "kinked"-edge silicon nanowires (edges in zigzag form) and demonstrated that the twisting and transverse phonon modes also suffer localization and reduce heat transport. Of course, localization is a coherent process, and inelastic scattering introduces decoherence, which could diminish localization. As Loh et al. [97] computed, this could be the case for optical phonons in graphene but not for acoustic modes due to larger inelastic MFPs for scattering compared to LLs. Since acoustic modes are the dominant heat carriers, the presence of disorder-caused localization will have a dramatic reduction in thermal transport. Below we compute the LLs of the different phonon modes in some of the GNRs we consider.

In the diffusive regime, the transmission decreases as $1/L$. In the localization regime, on the other hand, for channel lengths greater than the LL ($\zeta$), the transmission drops exponentially with a characteristic LL $\zeta$, as [101]

$$T_{\text{ph}}(\omega) \propto \exp\left[-\frac{L}{\zeta(\omega)}\right]. \tag{8.17}$$

Using a similar reasoning as in the extraction of the diffusive MFP for scattering, we extract the LL by

$$\zeta(\omega) = \frac{L_2 - L_1}{ln\left(\frac{T_{\text{LRS},L_1}(\omega)}{T_{\text{LRS},L_2}(\omega)}\right)}, \tag{8.18}$$

where it holds that $L_{1,2} >> \zeta(\omega)$.

In Fig. 8.9a, we also show the LL $\zeta(\omega)$ for the narrow $W = 1$ nm GNR (dashed blue line). To extract the LL we use Eq. 8.18, with $L_1 = 500$ nm and $L_2 = 1000$ nm. The LL features are very similar to the MFP features. Long LLs are observed at very low frequencies, reaching hundreds of nanometers. The LLs drop to a few nanometers for higher energies. Sharp dips are observed at energies $E \approx 0.16$ eV, which again correlates with the localized features in the $T \times L$ lines of Fig. 8.4f. In general, $\zeta(\omega)$ and $\lambda(\omega)$ are connected by the Thouless relation $\zeta(\omega)/\lambda(\omega) = N_m$ [102], where $N_{\text{m}}$ is the number of propagating modes in the pristine channel, in this case the same as the value of the ballistic transmission [101]. The ratio $\zeta(\omega)/\lambda(\omega)$ is shown in Fig. 8.9b for the $W = 1$ nm GNR (solid blue line), and as expected, it mostly follows the transmission trend (dashed black line).

We mention that dephasing mechanisms such as phonon-phonon scattering could prevent localization, which requires coherence. However, as the LL is in most of the spectrum smaller than the phonon-phonon scattering MFPs (see Ref. [57]), we expect that localization will be observed in this ultranarrow channel as described by the drop in $T \times L$ shown in Fig. 8.4f. Note that we do not attempt to compute the LLs for the wider $W = 5$ nm GNR. This is because from Fig. 8.4d it is obvious that modes from several parts of the spectrum are not localized at the channel lengths we were able to simulate. However, the large MFPs in this channel suggest even larger LLs, on the order of a few hundreds of nanometers. These lengths are similar to the dephasing lengths, or phonon-phonon scattering MFPs, as presented in Ref. [57], and therefore, localization could be prevented. On the other hand, introduction of a stronger LER amplitude on these wider GNRs would result in smaller roughness scattering MFPs and smaller LLs than the ones shown in Fig. 8.9a (red line). Smaller LLs could allow localization to appear, most probably at the same energies as they appear for the $W = 1$ nm GNR ($E \approx 0.073$ V, $E \approx 0.13$ eV, and $E \approx 0.16$ eV).

The important message to be conveyed from the calculations of $\lambda(\omega)$ and $\zeta(\omega)$ is that phonon transport in ultranarrow 1D channels consists of multiscale features, where phonons of MFPs from hundreds of nanometers down to a few nanometers are involved. Transport features can vary from ballistic to diffusive and to the localization regimes, depending on the phonon energy, level of disorder, channel length, and channel width. To properly understand phonon transport in 1D channels all of these features need to be taken into proper consideration.

## 8.7 Thermal Conductivity

Finally, it is important to extend the analysis to include features of thermal conductivity in ultranarrow GNRs. The thermal conductivity of the GNR channels is a length-dependent quantity and calculated using the thermal conductance as $\kappa_l = LK_l/A$, where $A$ is the cross-sectional area of the GNR, with its height assumed to be 0.335 nm. Figure 8.10 shows the thermal conductivity versus channel length for GNRs with width $W = 5$ nm (red crosses) down to $W = 1$ nm (blue circles). The increase in thermal conductivity with channel length for short channels and saturation for the longer ones indicates the transition between ballistic and diffusive transport, which was also observed at various instances [57]. For the wider GNR channels, the saturation begins for length scales of several hundreds of nanometers. At this channel length, however, the narrower GNR with $W = 1$ nm is already driven into the localization regime (blue line). Ballistic transport dictates that the thermal conductivity increases linearly with channel length, while saturation comes due to scattering. The strength of the LER is indicated by the deviation from unity of the slope (dashed line in Fig. 8.9) of the thermal conductivity lines for short channel lengths [103, 104]. A power-law behavior $L^\alpha$ is expected for 1D channels [103, 104]. From these calculations, for the wider channels $W = 4$ nm and 5 nm the slope is $\alpha = 0.7$. As the width decreases, the slope decreases as well, with the channel of $W = 3$ nm having $\alpha = 0.65$ and the narrowest channel, $W = 1$ nm, having $\alpha = 0.5$.

We note that although we consider freely suspended GNRs, our general conclusions would be valid for GNRs placed on substrates as well. Placing GNRs on a substrate introduces an additional scattering effect, namely substrate scattering, as described by Aksamija and Knezevic in Ref. [37]. In that work, Monte Carlo simulations for phonons in large-area GNRs on $SiO_2$ were carried out. The substrate scattering was modeled as a point interaction with small patches where the ribbon is in contact with the substrate [105]. The authors have shown that indeed substrate scattering is a dominant scattering mechanism, but the MFP of substrate scattering is about 67 nm, and it dominates phonon transport for GNRs with widths larger than 1 μm. For narrow GNRs with widths below $W = 130$ nm (the GNRs we investigate here are even smaller, down to a few nanometers), the LER dominates phonon transport. In any case, for substrate scattering to be modeled within NEGF, one should construct a dynamical matrix that could account for the deviation in bond length and angle. This should not be based on the FCM (in which case the parameters are constant) but potentially based on first-principle calculations or valence force fields (VFFs), in which the potential energy and force constants are calculated for each atom on the basis of its own position and the position of its neighbors. These methods require much larger computational time, in contrast to the FCM, however. The NEGF formalism could then be employed using that dynamical matrix.

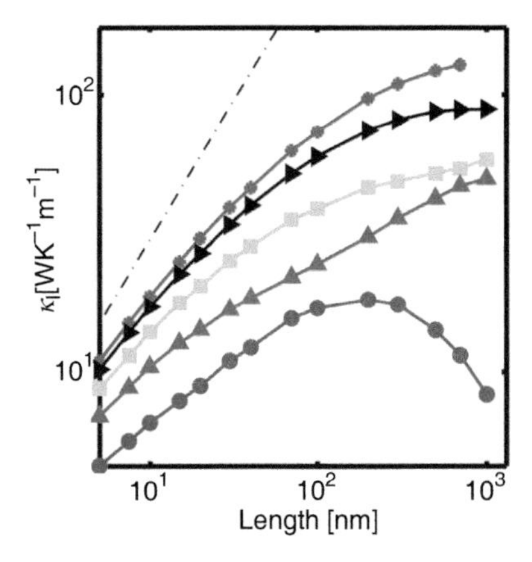


**Figure 8.10** Thermal conductivity versus channel length for GNRs with widths $W = 5$ nm (red crosses), $W = 4$ nm (black triangles), $W = 3$ nm (green squares), $W = 2$ nm (red triangles), and $W = 1$ nm (blue circles). The dashed line indicates the unit slope. Adapted figure with permission from Ref. [29]. Copyright (2015) by the American Physical Society.

## 8.8 Conclusions

In this chapter we presented an investigation of the thermal transport properties of low-dimensional, ultranarrow GNR channels under the influence of LER disorder. We employed the NEGF method for phonon transport and the FCM for the description of the phonon modes. We discussed that the LER affects different parts of the spectrum in different ways:

(i) Under strong effective disorder, thermal conductivity is dominated by low-frequency acoustic modes, which have MFPs of several hundred nanometers and suffer from localization only under extreme confinement in purely 1D channels. At ultranarrow channel widths they tend to completely dominate thermal transport.

(ii) Regions of the spectrum with a dense population of modes such as the optical modes can contribute significantly to thermal transport, even if their group velocity is low.

(iii) Regions of the spectrum with low mode density end up becoming effective transport gaps as the length of the channel increases or the width decreases and contribute little to thermal transport, even if they are relatively dispersive.

(iv) Regions of the spectrum with very low mode densities, populated with relatively flat modes, suffer from band mismatch in the presence of both confinement and roughness, which creates even stronger transport gaps and completely eliminates their ability to carry heat. In general, confinement reduces the population of the modes in the entire energy spectrum (except the low-frequency acoustic regions), and under the influence of disorder they fall into category (iii), that is, confinement and roughness reduce phonon transmission by introducing effective transport gaps and band mismatch. This drives transport at those energies into the localization regime.

Finally, we showed that although the transmission of several energy regions is severely degraded in the presence of the LER, for channels with lengths up to $L = 1$ μm that we have simulated, only the overall thermal conductivity of the ultranarrow $W = 1$ nm GNRs is driven into the localization regime.

## References


1. Mingo, N., and Broido, D. A. (2005). Length dependence of carbon nanotube thermal conductivity and the "problem of long waves," *Nano Lett.*, **5**, pp. 1221–1225.
2. Nika, D. L., Askerov, A. S., and Balandin, A. A. (2012). Anomalous size dependence of the thermal conductivity of graphene ribbons, *Nano Lett.*, **12**, pp. 3238–3244.
3. Chang, C. W., Okawa, D., Garcia, H., Majumdar, A., and Zettl, A. (2008). Breakdown of Fourier's law in nanotube thermal conductors, *Phys. Rev. Lett.*, **101**, p. 075903.
4. Lindsay, L., Broido, D. A., and Mingo, N. (2009). Lattice thermal conductivity of single-walled carbon nanotubes: beyond the relaxation time approximation and phonon-phonon scattering selection rules, *Phys. Rev. B*, **80**, p. 125407.
5. Lindsay, L., Broido, D. A., and Mingo N. (2010). Flexural phonons and thermal transport in graphene, *Phys. Rev. B*, **82**, p. 115427.
6. Nika, D. L., and Balandin, A. A. (2012). Two-dimensional phonon transport in graphene, *J. Phys.: Condens. Matter*, **24**, p. 233203.

7. Balandin, A. (2011). Thermal properties of graphene and nanostructured carbon materials, *Nat. Mater.*, **10**, pp. 569–581.
8. Xu, X., Pereira, L. F. C., Wang, Y., Wu, J., Zhang, K., Zhao, X., Bae, S., Bui, C. T., Xie, R., Thong, J. T. L., Hong, B. H., Loh, K. P., Donadio, D., Li, B., and Özyilmaz, B. (2014). Length-dependent thermal conductivity in suspended single-layer graphene, *Nat. Commun.*, **5**, p. 3689.
9. Savin, A. V., Kivshar, Y. S., and Hu, B. (2010). Suppression of thermal conductivity in graphene nanoribbons with rough edges, *Phys. Rev. B*, **82**, p. 195422.
10. Wang, Y., Qiu, B., and Ruan, X. (2012). Edge effect on thermal transport in graphene nanoribbons: a phonon localization mechanism beyond edge roughness scattering, *Appl. Phys. Lett.*, **101**, p. 013101.
11. Aksamija, Z., and Knezevic, I. (2014). Lattice thermal transport in large-area polycrystalline graphene, *Phys. Rev. B*, **90**, p. 035419.
12. Ni, X., Leek, M. L., Wang, J.-S., Feng, Y. P., and Li, B. (2011). Anomalous thermal transport in disordered harmonic chains and carbon nanotubes, *Phys. Rev. B*, **83**, p. 045408.
13. Lan, J., Wang, J.-S., Gan, C. K., and Chin, S. K. (2009). Edge effects on quantum thermal transport in graphene nanoribbons: tight-binding calculations, *Phys. Rev. B*, **79**, p. 115401.
14. Karamitaheri, H., Neophytou, N., Pourfath, M., Faez, R., and Kosina, H. (2012). Engineering enhanced thermoelectric properties in zigzag graphene nanoribbons, *J. Appl. Phys.*, **111**, p. 054501.
15. Mazzamuto, F., Saint-Martin, J., Valentin, A., Chassat, C., and Dollfus, P. (2011). Edge shape effect on vibrational modes in graphene nanoribbons: a numerical study, *J Appl. Phys.*, **109**, p. 064516.
16. Tan, Z. W., Wang, J.-S., and Gan, C. K. (2011). First-principles study of heat transport properties of graphene nanoribbons, *Nano Lett.*, **11**, pp. 214–219.
17. Hu, J., Ruan, X., and Chen, Y. P. (2009). Thermal conductivity and thermal rectification in graphene nanoribbons: a molecular dynamics study, *Nano Lett.*, **9**, pp. 2730–2735.
18. Fang, T., Konar, A., Xing, H., and Jena, D. (2008). Mobility in semiconducting graphene nanoribbons: phonon, impurity, and edge roughness scattering, *Phys. Rev. B*, **78**, p. 205403.
19. Fischetti, M. V., and Narayanan, S. (2011). An empirical pseudopotential approach to surface and line-edge roughness scattering in nanostructures: application to Si thin films and nanowires and to graphene nanoribbons, *J. Appl. Phys.*, **110**, p. 083713.

20. Fischetti, M. V., Kim, J., Narayanan, S., Ong, Z.-Y., Sachs, C., Ferry, D. K., and Aboud, S. J. (2013). Pseudopotential-based studies of electron transport in graphene and graphene nanoribbons, *J. Phys.: Condens. Matter*, **25**, p. 473202.
21. Takashima, K., and Yamamoto, T. (2014). Conductance fluctuation of edge-disordered graphene nanoribbons: crossover from diffusive transport to Anderson localization, *Appl. Phys. Lett.*, **104**, p. 093105.
22. Li, W., Sevinçli, H., Cuniberti, G., and Roche, S. (2010). Phonon transport in large scale carbon-based disordered materials: implementation of an efficient order-N and real-space Kubo methodology, *Phys. Rev. B*, **82**, p. 041410R.
23. Donadio, D., and Galli, G. (2009). Atomistic simulations of heat transport in silicon nanowires, *Phys. Rev. Lett.*, **102**, p. 195901.
24. Luisier, M. (2011). Investigation of thermal transport degradation in rough Si nanowires, *J. Appl. Phys.*, **110**, p. 074510.
25. Evans, W. J., Hu, L., and Keblinski, P. (2010). Thermal conductivity of graphene ribbons from equilibrium molecular dynamics: effect of ribbon width, edge roughness, and hydrogen termination, *Appl. Phys. Lett.*, **96**, p. 203112.
26. Karamitaheri, H., Pourfath, M., Faez, R., and Kosina, H. (2013). Atomistic study of the lattice thermal conductivity of rough graphene nanoribbons, *IEEE Trans. Electron Devices*, **60**, pp. 2142–2147.
27. Areshkin, D., Gunlycke, D., and White, C. (2007). Ballistic transport in graphene nanostrips in the presence of disorder: importance of edge effects, *Nano Lett.*, 7, pp. 204–210.
28. Neophytou, N., Ahmed, S., and Klimeck, G. (2007). Influence of vacancies on metallic nanotube transport properties, *Appl. Phys. Lett.*, **90**, p. 182119.
29. Karamitaheri, H., Pourfath, M., Kosina, H., and Neophytou, N. (2015). Low-dimensional phonon transport effects in ultranarrow disordered graphene nanoribbons, *Phys. Rev. B*, **91**, p. 165410.
30. Ghosh, S., Calizo, I., Teweldebrhan, D., Pokatilov, E. P., Nika, D. L., Balandin, A. A., Bao, W., Miao, F., and Lau, C. N. (2008). Extremely high thermal conductivity of graphene: prospects for thermal management applications in nanoelectronic circuits, *Appl. Phys. Lett.*, **92**, p. 151911.
31. Termentzidis, K., Barreteau, T., Ni, Y., Merabia, S., Zianni, X., Chalopin, Y., Chantrenne, P., and Voltz, S. (2013). Modulated SiC nanowires: molecular dynamics study of their thermal properties, *Phys. Rev. B*, **87**, p. 125410.

32. Melis, C., and Colombo, L. (2014). Lattice thermal conductivity of $Si_{1-x}Ge_x$ nanocomposites, *Phys. Rev. Lett.*, **112**, p. 065901.

33. Donadio, D., and Galli, G. (2010). Temperature dependence of the thermal conductivity of thin silicon nanowires, *Nano Lett.*, **10**, pp. 847–851.

34. Fthenakis, Z. G., Zhu, Z., and Tomanek, D. (2014). Effect of structural defects on the thermal conductivity of graphene: from point to line defects to haeckelites, *Phys. Rev. B*, **89**, p. 125421.

35. McGaughey, A. J. H., and Kaviany, M. (2006). Phonon transport in molecular dynamics simulations: formulation and thermal conductivity prediction, in *Advances in Heat Transfer*, Vol. 39, pp. 169–255, edited by Greene, G. A., Cho, Y. I., Hartnett, J. P., and Bar-Cohen, A., Elsevier, New York, NY.

36. Aksamija, Z., and Knezevic, I. (2010). Anisotropy and boundary scattering in the lattice thermal conductivity of silicon nanomembranes, *Phys. Rev. B*, **82**, p. 045319.

37. Aksamija, Z., and Knezevic, I. (2012). Thermal transport in graphene nanoribbons supported on $SiO_2$, *Phys. Rev. B*, **86**, p. 165426.

38. Wolf, S., Neophytou, N., and Kosina, H. (2014). Thermal conductivity of silicon nanomeshes: effects of porosity and roughness, *J. Appl. Phys.*, **115**, p. 204306.

39. Maldovan, M. (2012). Thermal conductivity of semiconductor nanowires from micro to nano length scales, *J. Appl. Phys.*, **111**, p. 024311.

40. McGaughey, A. J. H., and Kaviany, M. (2004). Quantitative validation of the Boltzmann transport equation phonon thermal conductivity model under the single-mode relaxation time approximation, *Phys. Rev. B*, **69**, p. 094303.

41. Ziman, J. M. (1960). *Electrons and Phonons*, Oxford University Press, London, UK.

42. Klemens, P. G. (1958). In *Solid State Physics*, Vol. 7, edited by Seitz, F., and Turnbull, D., Academic Press, New York, NY.

43. Huang, Z., Fisher, T. S., and Murthy, J. Y. (2010). Simulation of phonon transmission through graphene and graphene nanoribbons with a Green's function method, *J. Appl. Phys.*, **108**, p. 094319.

44. Huang, Z., Fisher, T. S., and Murthy, J. Y. (2010). Simulation of thermal conductance across dimensionally mismatched graphene interfaces, *J. Appl. Phys.*, **108**, p. 114310.

45. Xu, Y., Wang, J.-S., Duan, W., Gu, B.-L., and Li, B. (2008). Nonequilibrium Green's function method for phonon-phonon interactions and ballistic-diffusive thermal transport, *Phys. Rev. B*, **78**, p. 224303.

46. Yamamoto, T., and Watanabe, K. (2006). Nonequilibrium Green's function approach to phonon transport in defective carbon nanotubes, *Phys. Rev. Lett.*, **96**, p. 255503.

47. Mingo, N. (2006). Anharmonic phonon flow through molecular-sized junctions, *Phys. Rev. B*, **74**, p. 125402.

48. Markussen, T. (2012). Surface disordered Ge–Si core–shell nanowires as efficient thermoelectric materials, *Nano Lett.*, **12**, pp. 4698–4704.

49. Jeong, C., Datta, S., and Lundstrom, M. (2012). Thermal conductivity of bulk and thin-film silicon: a Landauer approach, *J. Appl. Phys.*, **111**, p. 093708.

50. Rego, L. G. C., and Kirczenow, G. (1998). Quantized thermal conductance of dielectric quantum wires, *Phys. Rev. Lett.*, **81**, p. 232.

51. Jeong, C., Datta, S., and Lundstrom, M. (2011). Full dispersion versus Debye model evaluation of lattice thermal conductivity with a Landauer approach, *J. Appl. Phys.*, **109**, p. 073718.

52. Wen, Y.-C., Hsieh, C.-L., Lin, K.-H., Chen, H.-P., Chin, S.-C., Hsiao, C.-L., Lin, Y.-T., Chang, C.-S., Chang, Y.-C., Tu, L.-W., and Sun, C.-K. (2009). Specular scattering probability of acoustic phonons in atomically flat interfaces, *Phys. Rev. Lett.*, **103**, p. 264301.

53. Broschat, S. L., and Thorsos, E. I. (1997). An investigation of the small slope approximation for scattering from rough surfaces. Part II. Numerical studies, *J. Acoust. Soc. Am.*, **101**, pp. 2615–2625.

54. Wang, M., Yang, N., and Guo, Z.-Y. (2011). Non-Fourier heat conductions in nanomaterials, *J. Appl. Phys.*, **110**, p. 064310.

55. Singh, D., Murthy, J. Y., and Fisher, T. S. (2011). On the accuracy of classical and long wavelength approximations for phonon transport in graphene, *J. Appl. Phys.*, **110**, p. 113510.

56. Ghosh, S., Bao, W., Nika, D. L., Subrina, S., Pokatilov, E. P., Lau, C. N., and Balandin, A. A. (2010). Dimensional crossover of thermal transport in few-layer graphene, *Nat. Mater.*, **9**, pp. 555–558.

57. Bae, M., Li, Z., Aksamija, Z., Martin, P. N., Xiong, F., Ong, Z., Knezevic, I., and Pop, E. (2013). Ballistic to diffusive crossover of heat flow in graphene ribbons, *Nat. Commun.*, **4**, p. 1734.

58. Karamitaheri, H., Neophytou, N., and Kosina, H. (2014). Anomalous diameter dependence of thermal transport in ultra-narrow Si nanowires, *J. Appl. Phys.*, **115**, p. 024302.

59. Lepri, S., Livi, R., and Politi, A. (1997). Heat conduction in chains of nonlinear oscillators, *Phys. Rev. Lett.*, **78**, p. 1896.

60. Li, B., and Wang, J. (2003). Anomalous heat conduction and anomalous diffusion in one-dimensional systems, *Phys. Rev. Lett.*, **91**, p. 044301.

61. Wu, G., and Dong, J. (2005). Anomalous heat conduction in a carbon nanowire: molecular dynamics calculations, *Phys. Rev. B*, **71**, p. 115410.

62. Hochbaum, A., Chen, R., Delgado, R., Liang, W., Garnett, E., Najarian, M., Majumdar, A., and Yang, P. (2008). Enhanced thermoelectric performance of rough silicon nanowires, *Nature*, **451**, pp. 163–167.

63. Boukai, A. I., Bunimovich, Y., Tahir-Kheli, J., Yu, J.-K., Goddard, W. A., and Heath, J. R. (2008). Silicon nanowires as efficient thermoelectric materials, *Nature*, **451**, pp. 168–171.

64. He, Y., Donadio, D., Lee, J.-H., Grossman, J. C., and Galli, G. (2011). Thermal transport in nanoporous silicon: interplay between disorder at mesoscopic and atomic scales, *ACS Nano*, **5**, pp. 1839–1844.

65. Venkatasubramanian, R. (2000). Lattice thermal conductivity reduction and phonon localizationlike behavior in superlattice structures, *Phys. Rev. B*, **61**, pp. 3091–3097.

66. Datta, S. (2005). *Quantum Transport: Atom to Transistor*, 2nd edn., Cambridge University Press, Cambridge, UK.

67. Jiang, J. W., Wang, B. S., and Wang, J. S. (2011). First principle study of the thermal conductance in graphene nanoribbon with vacancy and substitutional silicon defects, *Appl. Phys. Lett.*, **98**, p. 113114.

68. Ye, L. H., Liu, B. G., Wang, D. S., and Han, R. (2004). Ab initio phonon dispersions of single-wall carbon nanotubes, *Phys. Rev. B*, **69**, p. 235409.

69. Lobo, C., and Martins, J. (1997). Valence force field model for graphene and fullerenes, *Z. Phys. D*, **39**, p. 159.

70. Kusminskiy, S., Campbell, D., and Neto, A. C. (2009). Lenosky's energy and the phonon dispersion of graphene, *Phys. Rev. B*, **80**, p. 035401.

71. Wirtz, L., and Rubio, A. (2004). The phonon dispersion of graphite revisited, *Solid-State Commun.*, **131**, p. 141.

72. Wang, H., Wang, Y., Cao, X., Feng, M., and Lan, G. (2009). Vibrational properties of graphene and graphene layers, *J. Raman Spectrosc.*, **40**, p. 1791.

73. Karamitaheri, H., Neophytou, N., Pourfath, M., and Kosina, H. (2012). Study of thermal properties of graphene-based structures using the force constant method, *J. Comput. Electron.*, **11**, pp. 14–21.

74. Saito, R., Dresselhaus, M., and Dresselhaus, G. (1998). *Physical Properties of Carbon Nanotubes*, Imperial College Press, London, UK.

75. Mohr, M., Maultzsch, J., Dobardzic, E., Reich, S., Milosevic, I., Damnjanovic, M., Bosak, A., Krisch, M., and Thomsen, C. (2007). Phonon dispersion of graphite by inelastic x-ray scattering, *Phys. Rev. B*, **76**, p. 035439.

76. Shen, Y., Xie, G., Wei, X., Zhang, K., Tang, M., Zhong, J., Zhang, G., and Zhang, Y.-W. (2014). Size and boundary scattering controlled contribution of spectral phonons to the thermal conductivity in graphene ribbons, *J. Appl. Phys.*, **115**, p. 063507.

77. Lindsay, L., Li, W., Carrete, J., Mingo, N., Broido, D. A., and Reinecke, T. L. (2014). Phonon thermal transport in strained and unstrained graphene from first principles, *Phys. Rev. B*, **89**, p. 155426.

78. Pop, E., Varshney, V., and Roy, A. K. (2012). Thermal properties of graphene: fundamentals and applications, *MRS Bull.*, **37**, pp. 1273–1281.

79. Rowe, M. D. (2006). *Thermoelectrics Handbook: Macro to Nano*, Taylor and Francis Group, Boca Raton, FL.

80. Qian, J., Allen, M. J., Yang, Y., Dutta, M., and Stroscio, M. A. (2009). Quantized long-wavelength optical phonon modes in graphene nanoribbon in the elastic continuum model, *Superlattices Microstruct.*, **46**, p. 881.

81. Markussen, T., Jauho, A.-P., and Brandbyge, M. (2008). Heat conductance is strongly anisotropic for pristine silicon nanowires, *Nano Lett.*, **8**, pp. 3771–3775.

82. Yazdanpanah, A., Pourfath, M., Fathipour, M., Kosina, H., and Selberherr, S. (2012). A numerical study of line-edge roughness scattering in graphene nanoribbons, *IEEE Trans. Electron Devices*, **59**, pp. 433–440.

83. Scuracchio, P., Costamagna, S., Peeters, F. M., and Dobry, A. (2014). Role of atomic vacancies and boundary conditions on ballistic thermal transport in graphene nanoribbons, *Phys. Rev. B*, **90**, p. 035429.

84. Huang, H., Xu, Y., Zou, X., Wu, J., and Duan, W. (2013). Tuning thermal conduction via extended defects in graphene, *Phys. Rev. B*, **87**, p. 205415.

85. Vermeersch, B., Carrete, J., Mingo, N., and Shakouri, A. (2015). Superdiffusive heat conduction in semiconductor alloys. I. Theoretical foundations, *Phys. Rev. B*, **91**, p. 085202.

86. Vermeersch, B., Mohammed, A. M. S., Pernot, G., Koh, Y.-R., and Shakouri, A. (2015). Superdiffusive heat conduction in semiconductor alloys. II.

Truncated Lévy formalism for experimental analysis, *Phys. Rev. B*, **91**, p. 085203.

87. Gallo, N. A., and Molina, M. I. (2015). Bulk and surface bound states in the continuum, *J. Phys. A: Math. Theor.*, **48**, p. 045302.

88. Balaban, A. T., and Klein, D. J. (2009). Claromatic carbon nanostructures, *J. Phys. Chem. C*, **113**, p. 19123.

89. Aksamija, Z., and Knezevic, I. (2011). Lattice thermal conductivity of graphene nanoribbons: anisotropy and edge roughness scattering, *Appl. Phys. Lett.*, **98**, p. 141919.

90. Piscanec, S., Lazzeri, M., Mauri, F., Ferrari, A. C., and Robertson, J. (2004). Kohn anomalies and electron-phonon interactions in graphite, *Phys. Rev. Lett.*, **93**, p. 185503.

91. Lazzeri, M., Attaccalite, C., Wirtz, L., and Mauri, F. (2008). Impact of the electron-electron correlation on phonon dispersion: failure of LDA and GGA DFT functionals in graphene and graphite, *Phys. Rev. B*, **78**, p. 081406(R).

92. Baroni, S., de Gironcoli, S., and Corso, A. D. (2001). Phonons and related crystal properties from density-functional perturbation theory, *Rev. Mod. Phys.*, **73**, p. 515.

93. Zhou, J., and Dong, J. (2007). Vibrational property and Raman spectrum of carbon nanoribbon, *Appl. Phys. Lett.*, **91**, p. 173108.

94. Munoz, E., Lu, J., and Yakobson, B. I. (2010). Ballistic thermal conductance of graphene ribbons, *Nano Lett.*, **10**, pp. 1652–1656.

95. Hepplestone, S. P., and Srivastava, G. P. (2007). Low-temperature mean-free path of phonons in carbon nanotubes, *J. Phys. Conf. Ser.*, **92**, p. 012076.

96. Che, J. W., Cagin, T., and Goddard, W. A. (2000). Thermal conductivity of carbonnanotubes, *Nanotechnology*, **11**, pp. 65–69.

97. Loh, G. T., Teo, E. H. T., and Tay, B. K. (2012). Phonon localization around vacancies in graphene nanoribbons, *Diamond Relat. Mater.*, **23**, pp. 88–92.

98. Rodriguez-Nieva, J. F., Saito, R., Costa, S. D., and Dresselhaus, M. S. (2012). Effect of 13C isotope doping on the optical phonon modes in graphene: localization and Raman spectroscopy, *Phys. Rev. B*, **85**, p. 245406.

99. Kim, H. W., Ko, W., Ku, J. Y., Jeon, I., Kim, D., Kwon, H., Oh, Y., Ryu, S., Kuk, Y., Hwang, S. W., and Suh, H. (2015). Nanoscale control of phonon excitations in graphene, *Nat. Commun.*, **6**, p. 7528.

100. Jiang, J.-W., Yang, N., Wang, B.-S., and Rabczuk, T. (2013). Modulation of thermal conductivity in kinked silicon nanowires: phonon interchanging and pinching effects, *Nano Lett.*, **13**, pp. 1670–1674.

101. Savic, I., Mingo, N., and Stewart, D. A. (2008). Phonon transport in isotope-disordered carbon and boron-nitride nanotubes: is localization observable? *Phys. Rev. Lett.*, **101**, p. 165502.

102. Thouless, D. J. (1977). Maximum metallic resistance in thin wires, *Phys. Rev. Lett.*, **39**, p. 1167.

103. Guo, Z., Zhang, D., and Gong, X.-G. (2009). Thermal conductivity of graphene nanoribbons, *Appl. Phys. Lett.*, **95**, p. 163103.

104. Park, M., Lee, S.C., and Kim, Y. S. (2013). Length-dependent lattice thermal conductivity of graphene and its macroscopic limit, *J. Appl. Phys.*, **114**, p. 053506.

105. Seol, J. H., Jo, I., Moore, A. L., Lindsay, L., Aitken, Z. H., Pettes, M. T., Li, X., Yao, Z., Huang, R., Broido, D., Mingo, N., Ruoff, R. S., and Shi, L. (2010). Two-dimensional phonon transport in supported graphene, *Science*, **328**, p. 213.

# Index

9789814774413